OCEAN
ACIDIFICATION

A NATIONAL STRATEGY TO MEET THE CHALLENGES OF A CHANGING OCEAN

Committee on the Development of an Integrated Science Strategy for
Ocean Acidification Monitoring, Research, and Impacts Assessment

Ocean Studies Board

Division on Earth and Life Studies

NATIONAL RESEARCH COUNCIL
OF THE NATIONAL ACADEMIES

THE NATIONAL ACADEMIES PRESS
Washington, D.C.
www.nap.edu

THE NATIONAL ACADEMIES PRESS   500 Fifth Street, N.W.   Washington, DC 20001

NOTICE: The project that is the subject of this report was approved by the Governing Board of the National Research Council, whose members are drawn from the councils of the National Academy of Sciences, the National Academy of Engineering, and the Institute of Medicine. The members of the committee responsible for the report were chosen for their special competences and with regard for appropriate balance.

This study was supported by Contract/Grant No. DG133R-08-CQ-0062, OCE-0946330, NNX09AU42G, and G09AP00160 between the National Academy of Sciences and the National Oceanic and Atmospheric Administration, National Science Foundation, National Aeronautics and Space Administration, and U.S. Geological Survey. Any opinions, findings, conclusions, or recommendations expressed in this publication are those of the author(s) and do not necessarily reflect the views of the organizations or agencies that provided support for the project.

International Standard Book Number-13: 978-0-309-15359-1 (Book)
International Standard Book Number-10: 0-309-153659-X (Book)
International Standard Book Number-13: 978-0-309-15360-7 (PDF)
International Standard Book Number-10: 0-309-15360-X (PDF)
Library of Congress Catalog Card Number 2010934135

Additional copies of this report are available from the National Academies Press, 500 Fifth Street, N.W., Lockbox 285, Washington, DC 20055; (800) 624-6242 or (202) 334-3313 (in the Washington metropolitan area); Internet, http://www.nap.edu.

THE NATIONAL ACADEMIES
Advisers to the Nation on Science, Engineering, and Medicine

The **National Academy of Sciences** is a private, nonprofit, self-perpetuating society of distinguished scholars engaged in scientific and engineering research, dedicated to the furtherance of science and technology and to their use for the general welfare. Upon the authority of the charter granted to it by the Congress in 1863, the Academy has a mandate that requires it to advise the federal government on scientific and technical matters. Dr. Ralph J. Cicerone is president of the National Academy of Sciences.

The **National Academy of Engineering** was established in 1964, under the charter of the National Academy of Sciences, as a parallel organization of outstanding engineers. It is autonomous in its administration and in the selection of its members, sharing with the National Academy of Sciences the responsibility for advising the federal government. The National Academy of Engineering also sponsors engineering programs aimed at meeting national needs, encourages education and research, and recognizes the superior achievements of engineers. Dr. Charles M. Vest is president of the National Academy of Engineering.

The **Institute of Medicine** was established in 1970 by the National Academy of Sciences to secure the services of eminent members of appropriate professions in the examination of policy matters pertaining to the health of the public. The Institute acts under the responsibility given to the National Academy of Sciences by its congressional charter to be an adviser to the federal government and, upon its own initiative, to identify issues of medical care, research, and education. Dr. Harvey V. Fineberg is president of the Institute of Medicine.

The **National Research Council** was organized by the National Academy of Sciences in 1916 to associate the broad community of science and technology with the Academy's purposes of furthering knowledge and advising the federal government. Functioning in accordance with general policies determined by the Academy, the Council has become the principal operating agency of both the National Academy of Sciences and the National Academy of Engineering in providing services to the government, the public, and the scientific and engineering communities. The Council is administered jointly by both Academies and the Institute of Medicine. Dr. Ralph J. Cicerone and Dr. Charles M. Vest are chair and vice chair, respectively, of the National Research Council.

www.national-academies.org

Acknowledgments

This report was greatly enhanced by the participants of the meeting held as part of this study. The committee would first like to acknowledge the efforts of those who gave presentations at meetings: Richard Feely (NOAA), Steve Murawski (NOAA), Julie Morris (NSF), Paula Bontempi (NASA), Kevin Summers (EPA), John Haines (USGS), Emily Pidgeon (Conservation International), Mike Sigler (NOAA), Chris Langdon (Oregon State University), Steve Gittings (NOAA), George Waldbusser (Chesapeake Biological Laboratory), Joseph Kunkel (University of Massachusetts-Amherst), Stephen Carpenter (University of Wisconsin), Tim Killeen (NSF), Jerry Miller (OSTP), Rick Spinrad (NOAA), Hugh Ducklow (Marine Biological Laboratory), Daniel Schrag (Harvard University), Kai Lee (Packard Foundation), and Rob Lempert (RAND). These talks helped set the stage for fruitful discussions in the closed sessions that followed.

The committee is also grateful to a number of people who provided important discussion and/or material for this report: Mitch Covington, BugWare Inc.; Jason Hall-Spencer, University of Plymouth, UK; Russ Hopcroft, University of Alaska, Fairbanks; Howard Spero, University of California, Davis; and Richard Zimmerman, Old Dominion University.

This report has been reviewed in draft form by individuals chosen for their diverse perspectives and technical expertise, in accordance with procedures approved by the NRC's Report Review Committee. The purpose of this independent review is to provide candid and critical comments that will assist the institution in making its published report as sound as possible and to ensure that the report meets institutional standards for objectivity, evidence, and responsiveness to the study charge. The review comments and draft manuscript remain confidential to protect the integrity of the deliberative process. We wish to thank the following individuals for their participation in their review of this report:

Edward A. Boyle, Massachusetts Institute of Technology, Cambridge
Ken Caldeira, Carnegie Institution of Washington, Stanford, California
Stephen Carpenter, University of Wisconsin, Madison
Paul Falkowski, Rutgers University, New Brunswick, New Jersey
Jean-Pierre Gattuso, CNRS and Université Pierre et Marie Curie, France
Burke Hales, Oregon State University, Corvallis
David Karl, University of Hawaii, Honolulu
Chris Langdon, University of Miami, Florida
Paul Marshall, Great Barrier Reef Marine Park Authority, Queensland, Australia
Edward Miles, University of Washington, Seattle
Hans-Otto Pörtner, Alfred Wegener Institute, Bremerhaven, Germany
Andy Ridgewell, University of Bristol, United Kingdom
James Sanchirico, University of California, Davis
Brad Seibel, University of Rhode Island, Kingston

Although the reviewers listed above have provided many constructive comments and suggestions, they were not asked to endorse the conclusions or recommendations nor did they see the final draft of the report before its release. The review of this report was overseen by **Kenneth H. Brink**, Woods Hole Oceanographic Institution, appointed by the Division on Earth and Life Studies, and **W.L. Chameides**, Duke University, appointed by the Report Review Committee, who were responsible for making certain that an independent examination of this report was carried out in accordance with institutional procedures and that all review comments were carefully considered. Responsibility for the final content of this report rests entirely with the authoring committee and the institution.

Contents

Summary

The ocean absorbs a significant portion of carbon dioxide (CO_2) emissions from human activities, equivalent to about one-third of the total emissions for the past 200 years from fossil fuel combustion, cement production and land-use change (Sabine et al., 2004). Uptake of CO_2 by the ocean benefits society by moderating the rate of climate change but also causes unprecedented changes to ocean chemistry, decreasing the pH of the water and leading to a suite of chemical changes collectively known as ocean acidification. Like climate change, ocean acidification is a growing global problem that will intensify with continued CO_2 emissions and has the potential to change marine ecosystems and affect benefits to society.

The average pH of ocean surface waters has decreased by about 0.1 unit—from about 8.2 to 8.1—since the beginning of the industrial revolution, with model projections showing an additional 0.2-0.3 drop by the end of the century, even under optimistic scenarios (Caldeira and Wickett, 2005).[1] Perhaps more important is that the rate of this change exceeds any known change in ocean chemistry for at least 800,000 years (Ridgewell and Zeebe, 2005). The major changes in ocean chemistry caused by increasing atmospheric CO_2 are well understood and can be precisely calculated, despite some uncertainty resulting from biological feedback processes. However, the direct biological effects of ocean acidification are less certain

---

[1] "Acidification" does not mean that the ocean has a pH below neutrality. The average pH of the ocean is still basic (8.1), but because the pH is decreasing, it is described as undergoing acidification.

and will vary among organisms, with some coping well and others not at all. The long-term consequences of ocean acidification for marine biota are unknown, but changes in many ecosystems and the services they provide to society appear likely based on current understanding (Raven et al., 2005).

In response to these concerns, Congress requested that the National Research Council conduct a study on ocean acidification in the Magnuson-Stevens Fishery Conservation and Management Reauthorization Act of 2006. The *Committee on the Development of an Integrated Science Strategy for Ocean Acidification Monitoring, Research, and Impacts Assessment* is charged with reviewing the current state of knowledge and identifying key gaps in information to help federal agencies develop a program to improve understanding and address the consequences of ocean acidification (see Box S.1 for full statement of task). Shortly after the study was underway, Congress passed another law—the Federal Ocean Acidification Research and Monitoring (FOARAM) Act of 2009—which calls for, among other things, the establishment of a federal ocean acidification program; this report is directed to the ongoing strategic planning process for such a program.

Although ocean acidification research is in its infancy, there is already growing evidence of changes in ocean chemistry and ensuing biological impacts. Time-series measurements and other field data have documented the decrease in ocean pH and other related changes in seawater chemistry (Dore et al., 2009). The absorption of anthropogenic CO_2 by the oceans increases the concentration of hydrogen ions in seawater (quanti-

---

BOX S.1
Statement of Task

Among the many potential direct and indirect impacts of greenhouse gas emissions (particularly CO_2) and global warming, this study will examine the anticipated consequences of ocean acidification due to rising atmospheric carbon dioxide levels on fisheries, protected species, coral reefs, and other natural resources in the United States and internationally. The committee will recommend priorities for a national research, monitoring, and assessment plan to advance understanding of the biogeochemistry of carbon dioxide uptake in the ocean and the relationship to atmospheric levels of carbon dioxide, and to reduce uncertainties in projections of increasing ocean acidification and the potential effects on living marine resources and ocean ecosystems. The committee's report will:

continued

BOX S.1 Continued

1. Review current knowledge of ocean acidification, covering past, present, and anticipated future effects on ocean ecosystems.
    A. To what degree is the present understanding sufficient to guide federal and state agencies in evaluating potential impacts for environmental and living resource management?
    B. To what degree are federal agency programs and plans responsive to the nation's needs for ocean acidification research, monitoring and assessments?

2. Identify critical uncertainties and key science questions regarding the progression and impacts of ocean acidification and the new information needed to facilitate research and decision making for potential mitigation and adaptation options.
    A. What are the critical information requirements for impact assessments and forecasts (e.g., biogeochemical processes regulating atmospheric CO_2 exchange, buffering, and acidification; effects of acidification on organisms at various life stages and on biomineralization; and the effects of parallel stressors)?
    B. What should be the priorities for research and monitoring to provide the necessary information for national and regional impact assessments for living marine resources and ocean ecosystems over the next decade?
    C. How should the adverse impacts of ocean acidification be measured and valued?
    D. How could additional research and modeling improve contingency planning for adaptive management of acidification impacts on marine ecosystems and resources?

3. Recommend a strategy of research, monitoring, and assessment for federal agencies, the scientific community, and other partners, including a strategy for developing a comprehensive, coordinated interagency program to address the high priority information needs.
    A. What linkages with states, non-governmental organizations, and the international science community are required?
    B. What is the appropriate balance among (a) short and long term research goals and (b) research, observations, modeling, and communication?
    C. What opportunities are available to collaborate with international programs, such as the Integrated Marine Biogeochemistry and Ecosystem Research (IMBER) and Surface Ocean Lower Atmosphere Study (SOLAS) projects, and non-U.S. programs, such as the European Project on Ocean Acidification (EPOCA)? What would be the value of coordinating U.S. efforts through international scientific organizations such as the Intergovernmental Oceanographic Commission (IOC), the International Council for Science Scientific Committee on Oceanic Research (SCOR), the World Climate Research Programme (WCRP), the International Council for the Exploration of the Sea (ICES), and the North Pacific Marine Science Organization (PICES)?

fied as a decrease in pH), and also brings increases in CO_2 and bicarbonate ion concentrations and decreases the carbonate ion concentration. These changes in the inorganic carbon and acid-base chemistry of seawater can affect physiological processes in marine organisms such as carbon fixation in photosynthesis, maintenance of physiological pH in internal fluids and tissues, or precipitation of carbonate minerals. Some of the strongest evidence of the potential impacts of ocean acidification on marine ecosystems comes from experiments on calcifying organisms; acidifying seawater to various extents has been shown to affect the formation and dissolution of calcium carbonate shells and skeletons in a range of marine organisms including reef-building corals, commercially important mollusks such as oysters and mussels, and many phytoplankton and zooplankton species that form the base of marine food webs.

It is important to note that the concentration of atmospheric CO_2 is rising too rapidly for natural, $CaCO_3$-cycle (calcium carbonate) processes to maintain the pH of the ocean. As a consequence, the average pH of the ocean will continue to decrease as the surface ocean absorbs more atmospheric CO_2. In contrast, atmospheric CO_2 increased over thousands of years during the glacial/interglacial cycles of the past 800,000 years, slow enough for the $CaCO_3$ cycle to compensate and maintain near constant pH (Hönisch et al., 2009). In the deeper geologic past—many millions of years ago—atmospheric CO_2 reached levels multiple times higher than present conditions, resulting in a tropical climate up to the high latitudes. The similarity of these deep past events to the current anthropogenic increase in atmospheric CO_2 is unclear because the time frames for CO_2 release are not well constrained. If CO_2 levels increased over thousands of years during these deep past events, the $CaCO_3$ cycle would have stabilized the ocean against changes in pH (Caldeira et al., 1999). Better reconstructions of the time frame of those hot house/ice house CO_2 perturbations and the environmental conditions that ensued will be necessary to determine whether the changes in marine ecosystems observed in the fossil record reflect an increased acidification of the paleo-ocean during that time.

Experimental reduction of seawater pH with CO_2 affects many biological processes, including calcification, photosynthesis, nutrient acquisition, growth, reproduction, and survival, depending upon the amount of acidification and the species tested (Orr et al., 2009). It is currently not known if and how various marine organisms will ultimately acclimate or adapt to the chemical changes resulting from acidification, but existing data suggest that there likely will be ecological winners and losers, leading to shifts in the composition and functioning of many marine ecosystems. It is also not known how these changes will interact with other environmental stressors such as climate change, overfishing, and pollution. Most importantly, despite the potential for socioeconomic impacts to occur in coral reef systems, aquaculture, fisheries, and other sectors, there is not currently enough information to assess these impacts, much less develop plans to mitigate or adapt to them.

CONCLUSION: The chemistry of the ocean is changing at an unprecedented rate and magnitude due to anthropogenic carbon dioxide emissions; the rate of change exceeds any known to have occurred for at least the past hundreds of thousands of years.

Unless anthropogenic CO_2 emissions are substantially curbed, or atmospheric CO_2 is controlled by some other means, the average pH of the ocean will continue to fall. Ocean acidification has demonstrated impacts on many marine organisms. While the ultimate consequences are still unknown, there is a risk of ecosystem changes that threaten coral reefs, fisheries, protected species, and other natural resources of value to society.

CONCLUSION: Given that ocean acidification is an emerging field of research, the committee finds that the federal government has taken initial steps to respond to the nation's long-term needs and that the national ocean acidification program currently in development is a positive move toward coordinating these efforts.

An ocean acidification program will require coordination at the international, national, regional, state, and local levels. Within the U.S. federal government, it will involve many of the greater than 20 agencies that are engaged in ocean science and resource management. To address the full scope of potential impacts, strong interactions among scientists in multiple fields and from various organizations will be required and two-way communication with stakeholders will be necessary. Ultimately, a successful program will have an approach that integrates basic science with decision support.

The growing concern over ocean acidification is demonstrated in the several workshops that have been convened on the subject, as well as scientific reviews and community statements (e.g., Raven et al., 2005; Doney et al., 2009; Kleypas et al., 2006; Fabry et al., 2008a; Orr et al., 2009; European Science Foundation, 2009). These reviews and reports present a community-based statement on the science of ocean acidification as well as steps needed to better understand and address it; they provide the groundwork for the committee's analysis.

CONCLUSION: The development of a National Ocean Acidification Program will be a complex undertaking, but legislation has laid the foundation, and a path forward has been articulated in numerous

reports that provide a strong basis for identifying future needs and priorities for understanding and responding to ocean acidification.

The committee's recommendations, presented below, include six key elements of a successful national ocean acidification program: (1) a robust observing network, (2) research to fulfill critical information needs, (3) assessments and support to provide relevant information to decision makers, (4) data management, (5) facilities and training of ocean acidification researchers, and (6) effective program planning and management.

OBSERVING NETWORK

Many publications have noted the critical need for long-term monitoring of ocean and climate to document and quantify changes, including ocean acidification, and that the current observation systems for monitoring these changes are insufficient. A global network of robust and sustained chemical and biological observations will be necessary to establish a baseline and to detect and predict changes attributable to acidification.

The first step in developing the observing network will be identification of the appropriate chemical and biological parameters to be measured by the network and ensuring data quality and consistency across space and time. There is widespread agreement on the chemical parameters (and methods and tools for measurement) for monitoring ocean acidification. Unlike the chemical parameters, there are no agreed upon metrics for biological variables. In part, this is because the field is young and in part because the biological effects of ocean acidification, from the cellular to the ecosystem level, are very complex. To account for this complexity, the program will need to monitor parameters that cover a range of organisms and ecosystems and support both laboratory-based and field research. The development of new tools and techniques, including novel autonomous sensors, would greatly improve the ability to make relevant chemical and biological measurements over space and time and will be necessary to identify and characterize essential biological indicators concerning the ecosystem consequences of ocean acidification. As critical biological indicators and metrics are identified, the Program will need to incorporate those measurements into the research plan, and thus, adaptability in response to developments in the field is a critical element of the monitoring program.

The next step in developing the observing network will be consideration of available resources. A number of existing sites and surveys could serve as a backbone for an ocean acidification observational network, but these existing sites were not designed to observe ocean acidification and thus do not provide adequate coverage or measurements of key parameters. The current system of observations would be improved by adding sites and measurements in ecosystems projected to be vulnerable to ocean acidification (e.g., coral reefs and polar regions) and areas of high variability (e.g., coastal regions). Two community-based reports (Fabry et al., 2008a; Feely et al., 2010) identify vulnerable ecosystems, measurement requirements, and other details for developing an ocean acidification observational network. Another important consideration is the sustainability of long-term observations, which remains a perpetual challenge but is critical given the gradual, cumulative, and long-lasting pressure of ocean acidification. Integrating the network of ocean acidification observations with other ocean observing systems will help to ensure sustainability of the acidification-specific observations.

CONCLUSION: The chemical parameters that should be measured as part of an ocean acidification observational network and the methods to make those measurements are well established.

RECOMMENDATION: The National Program should support a chemical monitoring program that includes measurements of temperature, salinity, oxygen, nutrients critical to primary production, and at least two of the following four carbon parameters: dissolved inorganic carbon, pCO_2, total alkalinity, and pH. To account for variability in these values with depth, measurements should be made not just in the surface layer, but with consideration for different depth zones of interest, such as the deep sea, the oxygen minimum zone, or in coastal areas that experience periodic or seasonal hypoxia.

CONCLUSION: Standardized, appropriate parameters for monitoring the biological effects of ocean acidification cannot be determined until more is known concerning the physiological responses and population consequences of ocean acidification across a wide range of taxa.

RECOMMENDATION: To incorporate findings from future research, the National Program should support an adaptive monitoring program to identify biological response variables specific to ocean acidification. In the meantime, measurements of general indicators of ecosystem change, such as primary productivity, should be supported as part of a program for assessing the effects of acidification. These measurements will also have value in assessing the effects of other long-term environmental stressors.

RECOMMENDATION: To ensure long-term continuity of data sets across investigators, locations, and time, the National Ocean Acidifica-

tion Program should support inter-calibration, standards development, and efforts to make methods of acquiring chemical and biological data clear and consistent. The Program should support the development of satellite, ship-based, and autonomous sensors, as well as other methods and technologies, as part of a network for observing ocean acidification and its impacts. As the field advances and a consensus emerges, the Program should support the identification and standardization of biological parameters for monitoring ocean acidification and its effects.

CONCLUSION: The existing observing networks are inadequate for the task of monitoring ocean acidification and its effects. However, these networks can be used as the backbone of a broader monitoring network.

RECOMMENDATION: The National Ocean Acidification Program should review existing and emergent observing networks to identify existing measurements, chemical and biological, that could become part of a comprehensive ocean acidification observing network and to identify any critical spatial or temporal gaps in the current capacity to monitor ocean acidification. The Program should work to fill these gaps by:

- ensuring that existing coastal and oceanic carbon observing sites adequately measure the seawater carbonate system and a range of biological parameters;
- identifying and leveraging other long-term ocean monitoring programs by adding relevant chemical and biological measurements at existing and new sites;
- adding additional time-series sites, repeat transects, and in situ sensors in key areas that are currently undersampled. These should be prioritized based on ecological and societal vulnerabilities;
- deploying and field testing new remote sensing and in situ technologies for observing ocean acidification and its impacts; and
- supporting the development and application of new data analysis and modeling techniques for integrating satellite, ship-based, and in situ observations.

RECOMMENDATION: The National Ocean Acidification Program should plan for the long-term sustainability of an integrated ocean acidification observation network.

RESEARCH PRIORITIES

Ocean acidification research is still in its infancy. A great deal of research has been conducted and new information gathered in the past several years, and it is clear from this research that ocean acidification may threaten marine ecosystems and the services they provide. However, much more information is needed in order to fully understand and address these changes. Most previous research on the biological effects of ocean acidification has dealt with acute responses in a few species, and very little is known about the impacts of acidification on many ecologically or economically important organisms, their populations, and communities; the effects on a variety of physiological and biogeochemical processes; and the capacity of organisms to adapt to projected changes in ocean chemistry (Boyd et al., 2008). There is a need for research that provides a mechanistic understanding of physiological effects, elucidates the acclimation and adaptation potential of organisms, and allows scaling up to ecosystem effects, taking into account the role and response of humans in those systems and how best to support decision making in affected systems. There is also a need to understand these effects in light of multiple and potentially compounding environmental stressors, such as increasing temperature, pollution, and overfishing. The committee identifies eight broad research areas that address these critical information gaps; detailed research recommendations on specific regions and topics are contained in other community-based reports (i.e., Raven et al., 2005; Kleypas et al., 2006; Fabry et al., 2008a; Orr et al., 2009; Joint et al., 2009).

CONCLUSION: Present knowledge is insufficient to guide federal and state agencies in evaluating potential impacts for management purposes.

RECOMMENDATION: Federal and federally funded research on ocean acidification should focus on the following eight unranked priorities:

- **understand processes affecting acidification in coastal waters;**
- **understand the physiological mechanisms of biological responses;**
- **assess the potential for acclimation and adaptation;**
- **investigate the response of individuals, populations, and communities;**
- **understand ecosystem-level consequences;**
- **investigate the interactive effects of multiple stressors;**
- **understand the implications for biogeochemical cycles; and**
- **understand the socioeconomic impacts and inform decisions.**

ASSESSMENT AND DECISION SUPPORT

The FOARAM Act of 2009 charges an interagency working group with overseeing the development of impacts assessments and adaptation and mitigation strategies, and with facilitating communication and outreach with stakeholders. Because ocean acidification is a relatively new concern and research results are just emerging, it will be challenging to move from science to decision support. Nonetheless, ocean acidification is occurring now and will continue for some time. Resource managers will need information in order to adapt to changes in ocean chemistry and biology. In view of the limited current knowledge about the impacts of ocean acidification, the first step for the National Ocean Acidification Program will be to clearly define the problem and the stakeholders (i.e., for whom is this a problem and at what time scales), and build a process for decision support. It must be noted that a one-time identification of stakeholders and their concerns will not be adequate in the long term, and it should be considered an iterative process. As research is performed and the effects of ocean acidification are better defined, additional stakeholders may be identified, and the results of the socioeconomic analysis may change. For climate change decision support, there have been pilot programs within some federal agencies and there is growing interest within the federal government for developing a national climate service to further develop climate-related decision support. Similarly, new approaches for ecosystem-based management and marine spatial planning are also being developed. The National Ocean Acidification Program could leverage the expertise of these existing and future programs.

RECOMMENDATION: The National Ocean Acidification Program should focus on identifying, engaging, and responding to stakeholders in its assessment and decision support process and work with existing climate service and marine ecosystem management programs to develop a broad strategy for decision support.

DATA MANAGEMENT

Data quality and access, as well as appropriate standards for data reporting and archiving, will be integral components of a successful program to enhance the value of data collected and ensure they are accessible (with appropriate metadata) to researchers now and in the future. Other large-scale research programs have developed data policies that address data quality, access, and archiving to enhance the value of data collected within these programs, and the research community has developed *The Guide to Best Practices in Ocean Acidification Research and Data Reporting* to provide guidance on data reporting and usage (Riebesell et al., 2010).

A successful program will require a management office with sufficient resources to guide data management and synthesis, development of policies, and communication with principal investigators. There are many existing data management offices and databases that could support ocean acidification observational and research data.

The FOARAM Act also calls for an "Ocean Acidification Information Exchange" that would go beyond chemical and biological measurements alone, to produce syntheses and assessments that would be accessible to and understandable by managers, policy makers, and the general public. This is an important priority for decision support, but it would require specific resources and expertise, particularly in science communication, to operate effectively.

RECOMMENDATION: The National Ocean Acidification Program should create a data management office and provide it with adequate resources. Guided by experiences from previous and current large-scale research programs and the research community, the office should develop policies to ensure data and metadata quality, access, and archiving. The Program should identify appropriate data center(s) for archiving of ocean acidification data or, if existing data centers are inadequate, the Program should create its own.

RECOMMENDATION: In addition to management of research and observational data, the National Ocean Acidification Program, in establishing an Ocean Acidification Information Exchange, should provide timely research results, syntheses, and assessments that are of value to managers, policy makers, and the general public. The Program should develop a strategy and provide adequate resources for communication efforts.

FACILITIES AND HUMAN RESOURCES

Facilities and trained researchers will be needed to achieve the research priorities and observations described in this document. This may include large community resources and facilities including, for example, central facilities for high-quality carbonate chemistry measurements or technically complex experimental systems (e.g., free-ocean CO_2 experiment (FOCE)-type sites, mesocosms), facilities located at sites with natural pH gradients and variability, or intercomparison studies to enable integration of data from different investigators. There are some community facilities of this scale, but they are currently quite limited. Large facilities may be required to scale up to ecosystem-level experiments, although there are scientific and economic trade-offs among the various types of facilities.

Similarly, ocean acidification is a highly interdisciplinary and growing field that is attracting new graduate students, postdoctoral investigators, and principal investigators. Training opportunities to help scientists make the transition to this new field, and to engage researchers in fields related to management and decision support, will accelerate the progress in ocean acidification research.

RECOMMENDATION: As the National Ocean Acidification Program develops a research plan, the facilities and human resource needs should also be assessed. Existing community facilities available to support high-quality field- and laboratory-based carbonate chemistry measurements, well-controlled carbonate chemistry manipulations, and large-scale ecosystem manipulations and comparisons should be inventoried and gaps assessed based on research needs. An assessment should also be made of community data resources such as genome sequences for organisms vulnerable to ocean acidification. Where facilities or data resources are lacking, the Program should support their development, which in some cases also may require additional investments in technology development. The Program should also support the development of human resources through workshops, short-courses, or other training opportunities.

PROGRAM PLANNING, STRUCTURE, AND MANAGEMENT

The committee delineates ambitious priorities and goals for the National Ocean Acidification Program. The FOARAM Act calls for the development of a detailed, 10-year strategic plan for the National Ocean Acidification Program; while the ultimate details of such a plan are outside the scope of this report, the Program will need to lay out a clear strategic plan to identify key goals and set priorities, as well as a detailed implementation plan. Community input into plan development will promote transparency and community acceptance of the plans and Program. A 10-year plan allows for planned evaluations: in addition to a final 10-year assessment of the program, a mid-term review after 5 years would be useful in evaluating the progress toward the goals and making appropriate corrections. While the 10-year period outlined in the FOARAM Act may be adequate to achieve some goals, it is likely that the Program in its entirety will extend beyond this initial time frame and some operational elements may continue indefinitely. During the initial 10-year period, a legacy program for extended time series measurements, research, and management will need to be developed. The committee identifies eight key elements that will need to be included in the strategic plan (see below).

If fully executed, the elements outlined in the FOARAM Act and recommended in this report would create a large and complex program that will require sufficient support. These program goals are certainly on the order of, if not more ambitious than, previous major oceanographic programs and will require a high level of coordination that warrants a program office to coordinate the activities of the program and serve as a central point for communicating and collaborating with outside groups such as Congress and international ocean acidification programs. International collaboration is critical to the success of the Program; ocean acidification is a global problem which requires a multinational research approach. Such collaboration also affords opportunities to share resources (including expensive large-scale facilities for ecosystem-level manipulation) and expertise that may be beyond the capacity of one single nation.

RECOMMENDATION: The National Ocean Acidification Program should create a detailed implementation plan with community input. The plan should address (1) goals and objectives; (2) metrics for evaluation; (3) mechanisms for coordination, integration, and evaluation; (4) means to transition research and observational elements to operational status; (5) agency roles and responsibilities; (6) coordination with existing and developing national and international programs; (7) resource requirements; and (8) community input and external review.

RECOMMENDATION: The National Ocean Acidification Program should create a program office with the resources to ensure successful coordination and integration of all of the elements outlined in the FOARAM Act and this report.

1

Introduction

The oceans have absorbed a significant portion of all anthropogenic (CO_2) emissions (approximately a third of the CO_2 emitted from fossil fuel emissions, cement production and deforestation; Sabine et al., 2004), and in doing so have tempered the rise in atmospheric CO_2 levels and avoided some CO_2-related climate warming. In addition to playing a pivotal role in moderating climate, oceanic uptake of CO_2 is causing important changes in ocean chemistry and biology. Carbon dioxide dissolved in water acts as an acid, decreasing its pH,[1] and fostering a series of chemical changes. The entire process is known as ocean acidification.[2] Because it is another consequence of anthropogenic CO_2 emissions, ocean acidification has been dubbed "the other CO_2 problem" (Turley, 2005), and the "sleeper issue" (Freedman, 2008) of climate change. Ocean acidifica-

---

[1] The pH scale describes how acidic or basic a substance is, which is determined by the concentration of hydrogen ions (H^+). The scale ranges from 0 to 14, with 0 being highly acidic, 14 being highly basic, and 7 being neutral. Like the Richter scale, which measures earthquakes, the pH scale is logarithmic. Therefore, every unit on the pH scale represents a tenfold change in H^+ concentration. For example, the H^+ concentration at pH 4 is ten times more than at pH 5. Since preindustrial times, the pH of oceanic surface water has dropped from approximately 8.2 to 8.1; on a logarithmic scale, this approximately 0.1 unit change represents a 26% increase in the concentration of H^+ ions. There are different pH scales used by oceanographers; but the differences among them are small and not important in the context of this report.

[2] "Acidification" does not mean that the ocean has a pH below neutrality. The average pH of the ocean is still basic (8.1), but because the pH is decreasing, it is described as undergoing acidification.

tion, like climate change, is a growing problem that is linked to the rate and amount of CO_2 emissions and is expected to affect ecosystems and society on a global scale. Unlike the uncertainties regarding the extent of CO_2-induced climate change, the principal changes in seawater chemistry that result from an increase in CO_2 concentration can be measured or calculated precisely. Importantly, these chemical changes are also practically irreversible on a time scale of centuries due to the inherently slow turnover of biogeochemical cycles in the oceans.

The mean pH of the ocean's surface has decreased by about 0.1 unit (from approximately 8.2 to 8.1) since the beginning of the industrial revolution, representing a rate of change exceeding any known to have occurred for at least hundreds of thousands of years (Figure 1.1) (Raven et al., 2005). Model projections indicate that if emissions continue on their current trajectory (i.e., business-as-usual scenarios), pH may drop by another 0.3 units by the end of the century (e.g., Wolf-Gladrow et al., 1999; Caldeira and Wickett, 2003; Feely et al., 2004). Even under optimistic scenarios (i.e., SRES scenario B1[3]), mean ocean surface pH is expected to drop below 7.9 (e.g., Cooley and Doney, 2009).

Scientific research on the biological effects of acidification is still in its infancy and there is much uncertainty regarding its ultimate effects on marine ecosystems. But marine organisms will be affected by the chemical changes in their environment brought about by ocean acidification; the question is how and how much. A number of biological processes are already known to be sensitive to the foreseeable changes in seawater chemistry. A prime example is the impairment in the ability of some organisms to construct skeletons or protective structures made of calcium carbonate resulting from even a modest degree of acidification, although the underlying mechanisms responsible for this effect are not well understood. Effects on the physiology of individual organisms can be amplified through food web and other interactions, ultimately affecting entire ecosystems. Organisms forming oceanic ecosystems have evolved over millennia to an aqueous environment of remarkably constant composition. There is reason to be concerned about how they will acclimate or adapt to the changes resulting from ocean acidification—changes that are occurring very rapidly on geochemical and evolutionary time scales.

---

[3] The Intergovernmental Panel on Climate Change developed emissions projection scenarios by examining alternative development pathways that considered a wide range of demographic, economic, and technological drivers (IPCC, 2000).

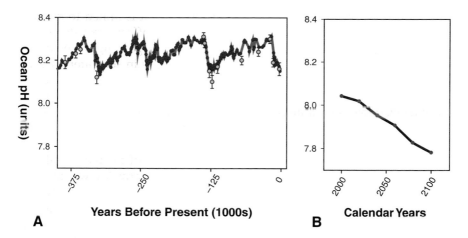

FIGURE 1.1 Estimated past, present, and future ocean pH (seawater scale). In panel A, past ocean pH was calculated from boron isotopes (see Box 2.2) in planktonic foraminifera shells (Hönisch et al., 2009, blue circles) and from ice core records of pCO_2, where alkalinity, salinity, and nutrients were assumed to remain constant (Petit et al., 1999, red circles). In panel B, the scale of the x-axis has been expanded to illustrate the pH trend projected over the next century. Future pH values (average for ocean surface waters) were calculated by assuming equilibrium with atmospheric pCO_2 levels and constant alkalinity. Future pCO_2 (atm) levels were assumed to follow the IS92 business-as-usual CO_2 emissions scenario.

1.1 CONTEXT FOR DECISION MAKING

It may seem that ocean acidification is a concern for the future. But ocean acidification is occurring now, and the urgent need for decision support is already quite evident. Recently, failures in oyster hatcheries in Oregon and Washington have been blamed on ocean acidification, and costly treatment systems have been installed, despite the fact that the evidence linking the failures to acidification is largely anecdotal (Welch, 2009). On the other hand, there is quite convincing evidence that coral reefs will be affected by acidification (see Chapter 4), but coral reef managers, who are just now beginning to develop adaptation plans to deal with climate change, have limited information on how to address acidification as well. These two examples highlight the urgent need for information on not only the consequences of acidification, but also how affected groups can adapt to these changes.

Like climate change, ocean acidification potentially affects governments, private organizations, and individuals—many of whom have insufficient information to consider fully the options for adaptation, mitigation, or policy development concerning the potentially far-reaching consequences of ocean acidification. While human activities have caused changes in the chemistry of the ocean in the past, none of those changes have been as fundamental, as widespread, and as long-lasting as those caused by ocean acidification. The resulting biological and ecological effects may not be as rapid and dramatic as those caused by other human activities (such as fishing and coastal pollution) but they will steadily increase over many years to come. Such long and gradual changes in ocean chemistry and biology—possibly punctuated by sudden ecological disruptions—undermines the foundation of existing empirical knowledge based on long-term studies of marine systems. Like climate change, ocean acidification renders past experience an undependable guide to decision making in the future.

To deal effectively with ocean acidification, decision makers will require new and different kinds of information and will need to develop new ways of thinking. For some, ocean acidification will be one more reason to reduce greenhouse gas emissions; for others, the priority will be coping with the ecological effects. But in all circumstances, more information to clarify, inform, and support choices will be needed. As is the case for climate change, decision support for ocean acidification will include "organized efforts to produce, disseminate, and facilitate the use of data and information in order to improve the quality and efficacy of (climate-related) decisions" (National Research Council, 2009a). The fundamental issue for ocean acidification decision support is the quality and timing of relevant information. Although the ongoing changes in ocean chemistry are well understood, the biological consequences are just now being elucidated. The problem is complicated because acidification is only one of a collection of stressful changes occurring in the world's oceans. It is also fundamentally difficult to understand how biological effects will cascade through food webs, and modify the structure and function of marine ecosystems. It may never be possible to predict with precision how and when acidification will affect a particular ecosystem. Ultimately, the information needed is related to social and economic impacts and pertain to "human dimensions" as has been noted in previous reports (e.g., National Research Council, 2008, 2009a). It is not only important to identify what user groups will be affected and when, but also to understand how resilient these groups are to the consequences of acidification and how capable they are of adapting to the changing circumstances.

To begin to address these societal concerns, the report tries to answer the questions of what to measure and why by identifying high-priority research and monitoring needs. It also addresses the process by identifying elements of an effective national strategy to help federal agencies provide the information needed by resource managers facing the impacts of ocean acidification in the marine environment.

1.2 STUDY ORIGIN AND POLICY CONTEXT

In the Magnuson-Stevens Fishery Conservation and Management Reauthorization Act of 2006 (P.L. 109-479, sec. 701), Congress called on "the Secretary of Commerce [to] request the National Research Council to conduct a study of the acidification of the oceans and how this process affects the United States." This request was reiterated in the Consolidated Appropriations Act of 2008 (P.L. 110-161). Based on these requests, the National Oceanic and Atmospheric Administration (NOAA) approached the Ocean Studies Board (OSB) to develop a study. While NOAA is a key federal agency in the effort to understand and address the consequences of ocean acidification, there are many other agencies involved in this topic. Therefore, NOAA and the OSB also sought input and sponsorship from the other members of the National Science and Technology Council Joint Subcommittee on Ocean Science and Technology (JSOST), composed of representatives from the 25 agencies that address ocean science and technology issues. JSOST assisted in developing the study terms and, in addition to NOAA, the National Science Foundation (NSF), the National Aeronautics and Space Administration (NASA), and the U.S. Geological Survey (USGS) agreed to support the study.

As the study was being developed, Congress enacted an additional law that would influence the committee's work. The Federal Ocean Acidification Research And Monitoring (FOARAM) Act of 2009 was passed as part of the Omnibus Public Land Management Act of 2009 (P.L. 111-11) and signed into law on March 30, 2009, shortly before the committee's first meeting. The purposes of the FOARAM Act are to:

- develop and coordinate an interagency plan for monitoring and research,
- establish an ocean acidification program within NOAA,
- assess and consider ecosystem and socioeconomic impacts, and
- research adaptation strategies and techniques for addressing ocean acidification.

The FOARAM Act outlines specific activities for both NOAA and NSF and also authorizes funds for these two agencies to carry out the Act, beginning at $14 million in fiscal year 2009 and ramping up to $35 million in 2012.

In light of this new law, the committee's work takes on added relevance. In parallel with the National Research Council (NRC) study, an interagency working group was assembled by the JSOST to develop the strategic plan. The committee considers this working group a primary audience for the report and hopes that the findings and recommendations feed into ongoing and future planning efforts by Congress and the federal agencies on ocean acidification research, monitoring, and impacts assessment.

1.3 STUDY APPROACH

The Committee on the Development of an Integrated Science Strategy for Ocean Acidification Monitoring, Research, and Impacts Assessment was assembled by the NRC to provide recommendations to the federal agencies on an interagency strategic plan for ocean acidification. The committee is charged with reviewing the current state of knowledge and identifying key gaps in information to ultimately help guide federal agencies with efforts to better understand and address the consequences of ocean acidification (see Box S.1 for full statement of task).

The committee recognizes that many thorough scientific reviews have already been published on the topic of ocean acidification (e.g., Raven et al., 2005; Fabry et al., 2008b; Doney et al., 2009). Rather than duplicate the previous work, the committee chose to focus on the issues most relevant to the interagency working group: the high priority information needs of decision makers and the key elements of an effective interagency program. The committee relied heavily on peer-reviewed literature, but also considered workshop reports, presentations at scientific meetings, and other community statements (e.g., Kleypas et al., 2006; Fabry et al., 2008a; Orr et al., 2009), as well as presentations at committee meetings and their own expert judgment as key inputs for establishing the community consensus on the current state of the science, research and monitoring priorities, and elements of an effective national program.

1.4 REPORT ORGANIZATION

The report begins with three chapters that provide a brief summary of the current knowledge on ocean acidification. Chapter 2 reviews the effects of increasing CO_2 concentration on seawater chemistry and discusses briefly what can be learned from the geological record, as well as possible mitigation options. Chapter 3 reviews what is known of the effects of acidification on the physiology of marine organisms. Chapter 4 addresses how these physiological affects may scale up and affect key marine ecosystems: tropical coral reefs, open ocean pelagic ecosystems, coastal margins, the deep sea (including cold-water corals), and high latitude ecosystems; it also includes a discussion of what may have occurred in the distant past and some general principles related to biodiversity and ecosystem thresholds. Chapter 5 addresses the evaluation and response to socioeconomic concerns of ocean acidification, with examples from three systems: fisheries, aquaculture, and tropical coral reefs. In Chapter 6, the committee lays out the groundwork for a national ocean acidification program.

2

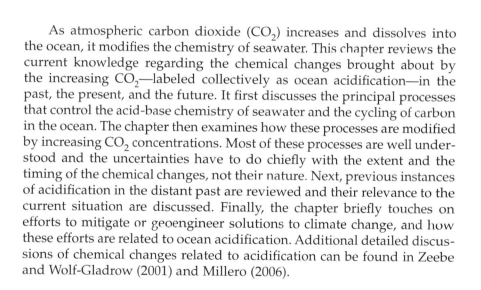

Effects of
Ocean Acidification
on the
Chemistry of Seawater

As atmospheric carbon dioxide (CO_2) increases and dissolves into the ocean, it modifies the chemistry of seawater. This chapter reviews the current knowledge regarding the chemical changes brought about by the increasing CO_2—labeled collectively as ocean acidification—in the past, the present, and the future. It first discusses the principal processes that control the acid-base chemistry of seawater and the cycling of carbon in the ocean. The chapter then examines how these processes are modified by increasing CO_2 concentrations. Most of these processes are well understood and the uncertainties have to do chiefly with the extent and the timing of the chemical changes, not their nature. Next, previous instances of acidification in the distant past are reviewed and their relevance to the current situation are discussed. Finally, the chapter briefly touches on efforts to mitigate or geoengineer solutions to climate change, and how these efforts are related to ocean acidification. Additional detailed discussions of chemical changes related to acidification can be found in Zeebe and Wolf-Gladrow (2001) and Millero (2006).

2.1 SEAWATER CHEMISTRY

The principal weak acids and bases that can exchange hydrogen ion in seawater and are thus responsible for controlling its pH are inorganic carbon species and, to a lesser extent, borate. Inorganic carbon dissolved in the ocean occurs in three principal forms: dissolved carbon dioxide

$(CO_2.aq)$,[1] bicarbonate ion (HCO_3^-), and carbonate ion (CO_3^{2-}) (see Box 2.1 for definitions.). CO_2 dissolved in seawater acts as an acid and provides hydrogen ions (H^+) to any added base to form bicarbonate:

$$CO_{2\,(aq)} + H_2O \rightleftharpoons H^+ + HCO_3^- \tag{1}$$

CO_3^{2-} acts as a base and takes up H^+ from any added acid to also form bicarbonate:

$$H^+ + CO_3^{2-} \rightleftharpoons HCO_3^- \tag{2}$$

Borate $[B(OH)_4^-]$ also acts as a base to take up H^+ from any acid to form boric acid $[B(OH)_3]$:

$$H^+ + B(OH)_4^- \rightleftharpoons B(OH)_3 + H_2O \tag{3}$$

As seen in reactions 1 and 2, bicarbonate can act as an acid or a base (i.e., donate or accept hydrogen ions) depending on conditions.

Under present-day conditions, these reactions buffer the pH of surface seawater at a slightly basic value of about 8.1 (above the neutral value around 7.0). At this pH, the total dissolved inorganic carbon (DIC ~ 2 mM) consists of approximately 1% CO_2, 90% HCO_3^-, and 9% CO_3^{2-} (Figure 2.1). The total boric acid concentration $(B(OH)_4^- + B(OH)_3)$ is about 1/5 that of DIC. As discussed in section 2.2, increases in CO_2 will increase the H^+ concentration, thus decreasing pH; the opposite occurs when CO_2 decreases. We note that isotope fractionation between $B(OH)_3$ and $B(OH)_4^-$ is used for estimating past pH values (Box 2.2).

Life in the oceans modifies the amount and forms (or species) of inorganic carbon and hence the acid-base chemistry of seawater. In the sunlit surface layer, phytoplankton convert, or "fix," CO_2 into organic matter during the day—a process also known as photosynthesis or primary production. This process simultaneously decreases DIC and increases the pH. The reverse occurs at night, when a portion of this organic matter is decomposed by a variety of organisms that regenerate CO_2, resulting in a daily cycle of pH in surface waters. A fraction of the particulate organic matter sinks below the surface where it is also decomposed, causing vertical variations in the concentrations of inorganic carbon species and pH. The net result is a characteristic maximum in CO_2 concentration and minima in pH and CO_3^{2-} concentration around 500 to 1,000 meters depth

---

[1] The proper notation for carbon dioxide gas is $CO_2.g$; carbon dioxide dissolved in water is $CO_2.aq$. However, for simplicity, these notations are not carried through the report; the text provides adequate context to determine which form of CO_2 is being discussed.

BOX 2.1
Parameters of the Ocean Acid-base System

DIC = Dissolved Inorganic Carbon concentration
DIC = $[CO_2] + [HCO_3^-] + [CO_3^{2-}]$
Where the brackets indicate concentrations in mol/Kg.

pCO_2 = partial pressure of CO_2 (in ppm or µatm)
$pCO_2 = [CO_2]/K_H$
Where K_H is the solubility constant for CO_2 in seawater (which varies with temperature, pressure and salinity)

Total Boric Acid = $[B(OH)_3] + [B(OH)_4^-]$

TA = Total Alkalinity
TA = $[HCO_3^-] + 2[CO_3^{2-}] + [B(OH)_4^-]$ + other minor bases

pH ≈ $-\log_{10} [H^+]$
More formally, oceanographers use two different pH scales, the total and the seawater pH scales:
$pH_T = -\log\{[H^+] + [HSO_4^-]\}$
$pH_{SWS} = -\log\{[H^+] + [HSO_4^-] + [HF]\}$
These two scales differ by about 0.01 units for a salinity S = 35 and temperature T = 25°C.

---

in many areas of the open ocean as illustrated in Figure 2.2a. Because the intensities of biological processes vary with season and the solubility of CO_2 varies with temperature, the pH and the concentrations of inorganic carbon species exhibit cyclical seasonal variations. For reasons discussed below, the vertical distribution of pH in the ocean varies with geographical location, particularly as a function of latitude; this is illustrated in the North-South transect for the Pacific Ocean in Figure 2.2b.

Another important process affecting the acid-base chemistry of seawater is the production of calcium carbonate ($CaCO_3$). Marine life produces the vast majority of $CaCO_3$ in the ocean; mostly in the form of the minerals calcite and aragonite (see Box 2.3). Even though these minerals are supersaturated in surface seawater, they do not normally precipitate spontaneously, but are formed by various organisms to serve as skeletons or hard protective structures. The degree of supersaturation of these minerals, quantified by the parameter Ω (see Box 2.3), varies with temperature, depth and seawater inorganic carbon chemistry; Ω is generally highest in shallow, warm waters and lowest in cold waters and at depth (Feely et al., 2004). When calcium carbonate sinks in the water column, it

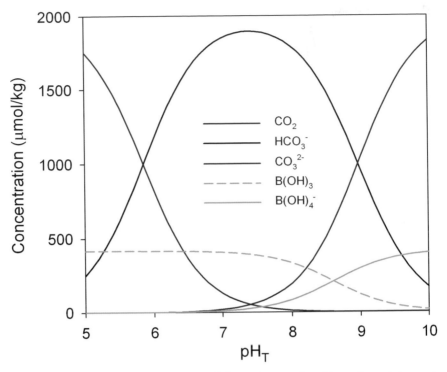

FIGURE 2.1 Typical concentrations of the major weak acids and weak bases in seawater as a function of pH. This diagram is calculated for constant dissolved inorganic carbon (DIC) and constant total boric acid using constants from Dickson et al. (2007) and Lueker et al. (2000).

BOX 2.2
Boron Isotopes as a Paleo-proxy for Seawater pH

Changes in ocean pH can be documented beyond the instrumental period of direct measurements using a proxy based on the incorporation into $CaCO_3$ of the borate ion, $B(OH)_4^-$ which has a lighter isotope composition than boric acid, $B(OH)_3$ (Spivack et al., 1993; Sanyal et al., 1995). For time scales shorter than the residence time of boron in the ocean—5-10 million years—measured values in sedimentary carbonates appear to accurately reflect the pH of the growth medium for several calcifying taxa. Results from glacial-interglacial times generally reflect the pH-buffering effect of the $CaCO_3$ cycle (Hönisch, 2005), while records from more recent time intervals reflect acidification of the ocean from rising CO_2 concentrations over the past centuries (Liu, 2009).

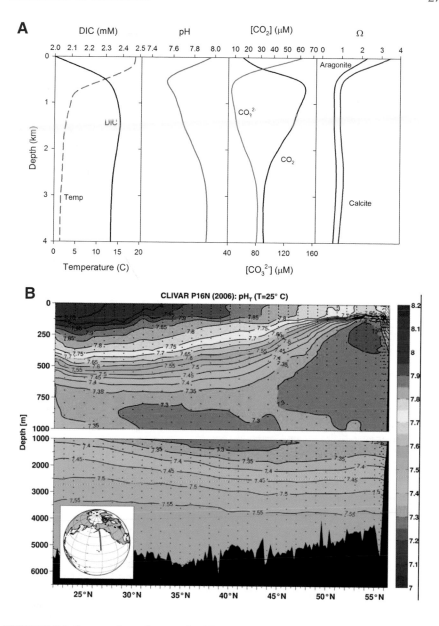

FIGURE 2.2 Inorganic carbon and pH vary as a function of depth and latitude. (a) Vertical profiles typical of the mid-North Pacific showing variations of several seawater chemical parameters with depth. Adapted from Morel and Hering (1993) with calculations using constants from Dickson et al. (2007) and Lueker et al. (2000). (b) Typical distribution of pH with depth along a North-South transect for the Pacific Ocean. (Byrne et al., 2010a).

BOX 2.3
Calcium Carbonate Solubility

Many marine organisms deposit calcareous shells and skeletons made of calcium carbonate ($CaCO_3$), which is a soluble mineral (Sanyal et al., 1995). The solubility of minerals such as $CaCO_3$ varies depending upon the physical properties of the seawater (e.g., temperature, salinity, and pressure) and also the crystal form of the mineral. The solubility is often expressed as the saturation state (Ω) of a mineral: when $\Omega > 1$, seawater is supersaturated with respect to $CaCO_3$ and it will remain solid; when $\Omega < 1$, seawater is undersaturated and $CaCO_3$ structures may begin to dissolve, unless they are protected from dissolution (e.g., with an organic coating). The saturation state is defined as follows:

$$\Omega = \frac{\left[Ca^{2+}\right]_{sw} \times \left[CO_3^{2-}\right]_{sw}}{\left[Ca^{2+}\right]_{sat} \times \left[CO_3^{2-}\right]_{sat}}$$

The denominator refers to the stoichiometric solubility product (often designated as K_{sp}) of the Ca^{2+} and CO_3^{2-} concentrations in a solution saturated with respect to the given mineral, and the numerator is the product of the in situ concentrations. Under current pH conditions, $CaCO_3$ is supersaturated in most surface ocean waters. Calcium ion concentration varies little in the open ocean, but ocean acidification decreases the concentration of CO_3^{2-} and the degree of supersaturation. In estuarine waters both Ca^{2+} and CO_3^{2-} concentrations vary widely and can frequently be below saturation.

Most calcium carbonate is precipitated by organisms in one of two forms: calcite (which has a rhombohedral crystal structure) and aragonite (which is orthorhombic). Vaterite, a third form, is rare but of interest because it is involved in the early stages of calcite precipitation in some organisms and is highly soluble. Normally, aragonite is about 1.5 times more soluble in seawater than calcite. However, the calcite crystal structure allows some ionic substitution of magnesium (Mg) for calcium: calcite with > 4 mol% $MgCO_3$ is called "high-Mg calcite" and is usually more soluble than regular calcite.

FROM: Morse and Mackenzie, 1990 and Morse et al., 2006.

---

becomes less stable (Ω decreases) as a result of the decrease in CO_3^{2-} concentration and the increase in the solubility of the minerals caused by the higher pressure and the lower temperature. The depth at which $CaCO_3$ becomes undersaturated and begins to dissolve depends on its crystalline form; this "saturation horizon" for calcite is deeper than that for aragonite (see Box 2.3). Precipitation of $CaCO_3$ at the surface lowers the ambient pH, while its dissolution at depth increases it, partially compensating for the inverse effects of the photosynthetic reduction of CO_2 that raises pH in surface waters and lowers pH in deeper waters as CO_2 is regenerated by metabolic oxidation.

As illustrated in Figure 2.2b, the vertical distribution of pH is not uniform throughout the oceans. The principal cause of these geographical pH variations is the non-uniform distribution of the CO_2 concentration resulting from the lower solubility of CO_2 gas at higher temperatures, basin-wide patterns of subsurface biological oxidation of organic matter and dissolution of carbonate minerals, and upwelling of CO_2 rich deep water or downwelling of CO_2-poor surface water (Sarmiento and Gruber, 2006). This is illustrated in Figure 2.3 (Part A) which shows CO_2 concentration as a function of depth in a North-South transect across the North Pacific Ocean. Upwelling around the equator increases CO_2 concentration near the surface at low latitudes compared to values in mid latitudes. An increase in surface CO_2 is also seen at high latitudes caused by the high solubility of CO_2 in cold water. High concentrations in deeper water result from oxidation of organic matter. These geographical patterns in CO_2 con-

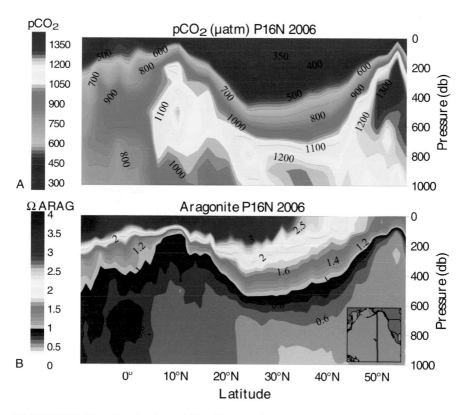

FIGURE 2.3 The distribution of (a) pCO_2 and (b) aragonite saturation in the North Pacific Ocean during a transect in March 2006. A pressure of 1 decibar (1 db on the y axis) corresponds approximately to a depth of 1 meter (m). (Fabry et al., 2008b)

centration are reflected in consistent patterns of CO_3^{2-} concentrations and thus also in the degree of saturation (Ω) of $CaCO_3$ minerals (see Figure 2.3, Part B) and in the buffering capacity of the water (Egleston et al., 2010).

2.2 ANTHROPOGENIC CARBON DIOXIDE EMISSIONS AND OCEAN ACIDIFICATION

The exchange of CO_2 at the air-water interface is relatively fast, taking place on a time scale of months to a year so that, on average, the concentration of CO_2 in surface seawater remains approximately at equilibrium with that of the atmosphere. As the concentration of atmospheric CO_2 gas increases year after year, some of it dissolves into the ocean such that about a third of the total CO_2 added to the atmosphere from anthropogenic sources—including fossil fuel emissions, cement production and deforestation—over the past 150 years is now dissolved in the oceans (Sabine et al., 2004; Khatiwala et al., 2009). The increase in dissolved CO_2 concentration decreases the pH and shifts the equilibrium of inorganic carbon species in seawater, resulting in an increase in CO_2 and HCO_3^- concentrations and a decrease in CO_3^{2-} concentration (Figure 2.4). For example, under present conditions in the mid North Pacific, for every 100 molecules of CO_2 dissolved from the atmosphere, about 7 remain as CO_2, 15 react with $B(OH)_4^-$, and 78 react with CO_3^{2-}, resulting in an increase of HCO_3^- by 171 molecules. The buffering capacity of seawater— the ability to resist changes in acid-base chemistry upon addition of an acid such as CO_2—depends on the concentration of bases, principally CO_3^{2-} and $B(OH)_4^-$, to neutralize the acid (Figures 2.1 and 2.4). Upon acidification of the oceans, the buffering capacity of seawater will decrease along with pH. Also, ocean water masses that are presently already high in CO_2 for any reason are less buffered against further increases in CO_2 than those with lower CO_2 (Egleston et al., 2010).

The decrease in carbonate ion concentration, CO_3^{2-}, that results from ocean acidification will lead to reduced rates of calcification, along with the a shoaling of the saturation horizons for calcium carbonate minerals to shallower depths, and a change in the marine calcium carbonate cycle. The resulting overall decrease in $CaCO_3$ precipitation and burial will tend to raise seawater pH, favoring the oceanic uptake of CO_2, and providing a small negative feedback on rising atmospheric CO_2 and global warming (Heinze, 2004). The extent of this feedback depends in part on the relative contributions of calcite and aragonite, and hence of the organisms that produce them, to the $CaCO_3$ cycle. Model simulations (Gehlen et al., 2007) show that an approximately 30% reduction in $CaCO_3$ production (which was hypothesized to occur when atmospheric CO_2 reached 4x pre-industrial values) leads to an additional cumulative oceanic uptake

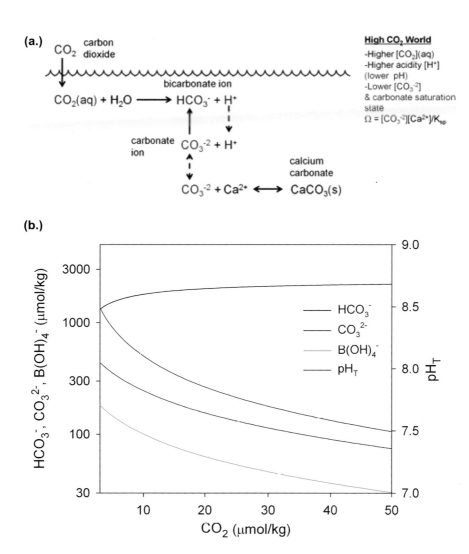

FIGURE 2.4 Schematic (a) and calculations (b) showing the effect of increasing CO_2 concentration on acid-base species in seawater. Calculations are made for constant alkalinity using constants from Dickson et al. (2007) and Lueker et al. (2000). Note that the y-axis is on log scale.

of ~6 petagrams (Pg) C, small relative to anthropogenic emissions and other potential climate-carbon cycle feedbacks (Friedlingstein et al., 2006). The reduction in carbonate production and its faster dissolution rate in the water column could also decrease the ballasting of organic carbon by $CaCO_3$ that increases the sinking of organic carbon to the deep ocean (e.g., Armstrong et al., 2002; Klaas and Archer, 2002). This would cause more organic carbon to decompose in shallow water and partially offset the negative CO_2 feedback resulting from lower calcification rates (Heinze, 2004). This effect could be enhanced by an increase in phytoplankton production of extracellular organic carbon (see chapters 3 and 4) and by the accelerated bacterial decomposition of organic matter at higher temperature.

A decrease in seawater pH results in a readjustment of all minor acid-base species, in addition to inorganic carbon and borate. These include a myriad of trace organic compounds, inorganic species such as the hydroxyl ion, phosphate and ammonium, and trace metals bound to inorganic or organic compounds. The effect of pH on these chemical species is of interest because several are important nutrients for phytoplankton growth and the chemical forms affect availability for phytoplankton use. For example, iron (Fe) is the most important trace nutrient for marine phytoplankton and inorganic Fe compounds are more biologically available than organically-bound Fe; acidification may cause Fe to become less bioavailable because as the pH decreases, more Fe will become organically bound (Shi et al., 2010). The effect of decreasing pH on Fe bioavailability in surface water is further complicated by the light-induced cycle between oxidized and reduced Fe species, in which a key process—oxidation of reduced Fe—slows down at lower pH. Such effects of acidification on the chemistry and bioavailability of trace metals and other compounds in the ocean have barely been studied at all and, unlike the changes in inorganic carbon species, cannot be predicted with confidence.

In addition, recent studies have shown that ocean acidification can affect the physical properties of seawater. At low frequencies, sound transmission in the ocean is attenuated by volume changes related to acid-base equilibrium of some chemical species. Change in the proportions of such systems, notably the boric acid and borate ion acid-base pair, may thus result in a "noisier ocean" (Hester et al., 2008; Duda, 2009).

2.2.1 Projections for Surface Waters

Because the relationship between atmospheric CO_2 and seawater carbonate chemistry is well understood, it is a simple matter to calculate the variations in average pH and inorganic carbon species concentrations in the surface waters of the open ocean based on the known variations in atmospheric CO_2 over the past 150 years (from actual measurements or from ice core data). Independent estimation of past seawater pH have been made using boron isotopes as well (see Box 2.2). Similarly, projections for changes in seawater chemistry can be made for the future on the basis of any future CO_2 emission scenario such as those published by the IPCC. Such calculations are shown in Figure 2.5 for the Pacific Ocean; models show that, based on a "business-as-usual" scenario of CO_2 emissions, the surface ocean pH will decrease by about 0.3 units within the next 100-150 years (e.g., Wolf-Gladrow et al., 1999; Caldeira and Wickett, 2003; Feely et al., 2004).

Figure 2.6 shows the results of actual measurements of surface seawater chemistry at a station near Hawaii between 1998 and 2008. These data confirm the validity of the calculations and demonstrate the predicted trend of a decrease of about 0.0015 pH units per year. The data also illustrate the seasonal cycle in pH and inorganic carbon species caused

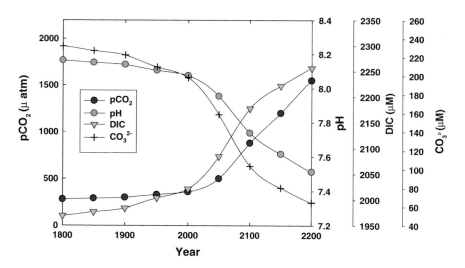

Expected Changes in the CO_2 System

FIGURE 2.5 Projected changes in the pH, and the concentrations of CO_2 and CO_3^{2-} in surface seawater under a business as usual scenario for CO_2 emissions over the next two centuries. Calculations were made for a salinity of 35 and temperature of 25°C assuming constant alkalinity using the CO_2sys program (Lewis and Wallace, 1998). The projected future values of pCO_2 in the atmosphere are based on the estimates of Caldeira and Wickett (2003).

by variations in biological activity discussed above. Because the buffering capacity of seawater decreases with decreasing pH, it is expected that these seasonal variations will amplify in the future.

2.2.2 Projections for Deeper Waters

While the CO_2 concentration in the surface ocean tracks the increasing values in the atmosphere, the penetration of that CO_2 into deep water depends on the slow vertical mixing of the water column and the transport of water masses in the complex wind-driven circulation and overturning of the oceans (Sarmiento and Gruber, 2006). About half of the anthropogenic CO_2 is now found in the upper 400 meters, while the other half has penetrated to deeper water, as illustrated in Figure 2.7 (Feely et al., 2004). This slow penetration of CO_2 into the deep ocean is reflected in a slower decrease in pH at depth than at the surface. An illustration of the time lag between surface and deep ocean acidification is shown in Figure 2.8; according to these simple calculations, under a "business-as-usual" scenario of CO_2 emissions, it will take about 500 years longer for a 0.3 unit decrease to occur in deep waters compared to surface waters (Caldeira and Wickett, 2003). However, in some regions where the vertical movement of water is relatively fast, the time scale for deep penetration of anthropogenic CO_2 will be on the order of decades instead of centuries (Sabine et al., 2004).

FIGURE 2.6 [*next page*] Time-series of mean carbonic acid system measurements within selected depth layers at Station ALOHA, 1988–2007. (First image) Partial pressure of CO_2 in seawater calculated from DIC and TA (blue symbols) and in water-saturated air at in situ seawater temperature (red symbols). Linear regressions of the sea and air pCO_2 values are represented by solid and dashed lines, respectively. (Second, third, and fourth images) In situ pH, based on direct measurements (orange symbols) or as calculated from DIC and TA (green symbols), in the surface layer and within layers centered at 250 and 1,000 m. Linear regressions of the calculated and measured pH values are represented by solid and dashed lines, respectively. (Dore et al., 2009)

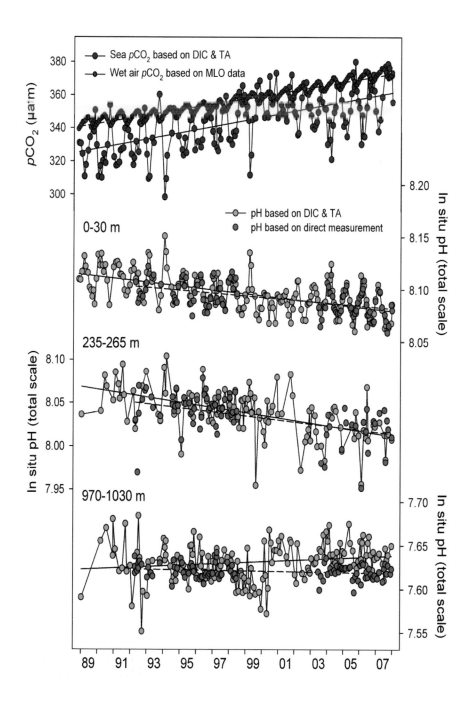

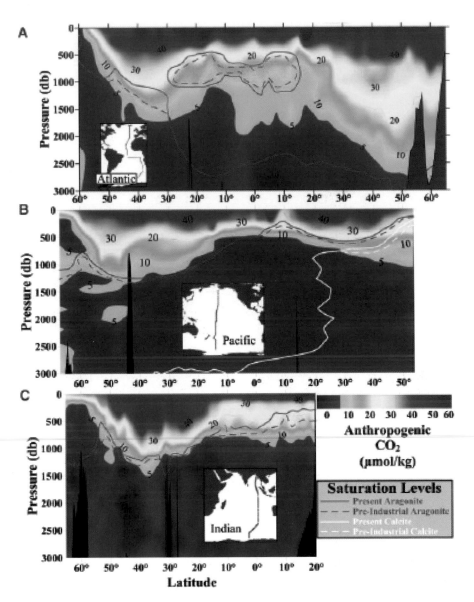

FIGURE 2.7 Vertical distributions of anthropogenic CO_2 concentrations (μmol kg^{-1}) and the saturation horizons for aragonite and calcite along north-south transects in the (A) Atlantic, (B) Pacific, and (C) Indian Oceans. A pressure of 1 decibar (1 db on the y-axis) corresponds approximately to a depth of 1 meter (m). (Feely et al., 2004)

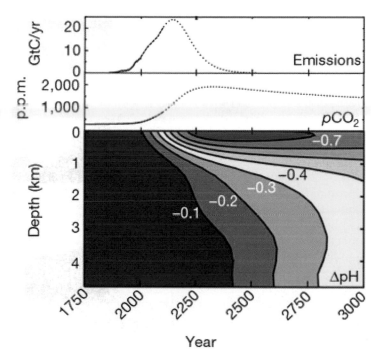

FIGURE 2.8 Atmospheric CO_2 emissions, historical atmospheric CO_2 levels and predicted CO_2 concentrations from this emissions scenario, together with changes in ocean pH based on horizontally averaged chemistry. (Caldeira and Wickett, 2003)

As anthropogenic CO_2 penetrates down in the water column, it decreases the CO_3^{2-} concentration and hence the degree of $CaCO_3$ supersaturation. The result is a slow upward migration, or shoaling, of the saturation horizons for calcite and aragonite. This effect can already be measured (Figure 2.7; Feely et al., 2004). As can be seen on Figure 2.7, the extent of shoaling of the saturation horizons is uneven across ocean basins, reflecting the differences in CO_2 penetration caused by the complex movements of water masses.

2.2.3 Projections for Coastal Waters

The acid-base chemistry of coastal waters is much more complex than that of open ocean surface and deep waters. It is affected by freshwater and atmospheric inputs, the supply of both organic matter and algal nutrients from land, and processes in the underlying sediments.

Fresh water runoff tends to have higher dissolved CO_2 concentrations and lower pH than ocean water (Salisbury et al., 2008). In surface coastal waters, high photosynthetic activity fueled by nutrient inputs can result in low seasonal CO_2 concentrations and high pH. In bottom waters, the decomposition of organic matter, contributed either from land or from local production, increases CO_2 and decreases pH. A number of anthropogenic activities can exacerbate coastal acidification, principally those that result in inputs of organic waste or algal nutrients, or that lead to the formation of acid rain (Doney et al., 2007).

Many coastal areas also experience seasonal upwelling of CO_2-rich deep water. In general, deep old waters in the ocean tend to have the least invasion of fossil fuel CO_2 but some upwelled waters are from shallower waters that are already subject to acidification by anthropogenic CO_2. This phenomenon has been shown to occur on the Pacific coast of North America (Figure 2.9; Feely et al., 2008). On that coast, the seasonal upwelling results in a natural seasonal cycle in pH and seawater carbonate chemistry; the extent and degree to which this has been amplified by acidification, resulting in the breaching of corrosive, aragonite dissolving water all the way to the surface, is an important research question. In both river dominated and upwelling dominated coastal regions, future trends in seawater carbon chemistry may also depend strongly on climate change that influences wind patterns, upwelling and river flow. In shallow waters, sediment dissolution can partly buffer acid inputs (Andersson et al., 2003; Thomas et al., 2009).

2.2.4 Projections for High Latitudes

As seen in Figure 2.3, the cold waters of high latitude regions are naturally low in carbonate ion concentration, owing to the increased solubility of CO_2 at low temperature and ocean mixing patterns. As a result, surface waters of these areas naturally have a lower degree of supersaturation of carbonate minerals and their acid-base chemistry is less buffered than temperate and tropical surface waters. As the atmospheric CO_2 concentration increases, the pH and CO_3^{2-} concentration in these regions will decrease, and the saturation horizons of aragonite and calcite will move rapidly toward the surface (Olafsson et al., 2009). Seasonal aragonite undersaturation in surface waters has already been observed in the Canada Basin of the Arctic Ocean (Bates et al., 2009; Yamamoto-Kawai et al., 2009). Persistent undersaturation of surface waters with respect to aragonite is projected to occur in high latitude regions by 2100, while in lower latitude surface waters the degree or extent of supersaturation will be reduced (Orr et al., 2005; Steinacher et al., 2009). This is illustrated in Figure 2.10, which shows the projected changes in the aragonite satura-

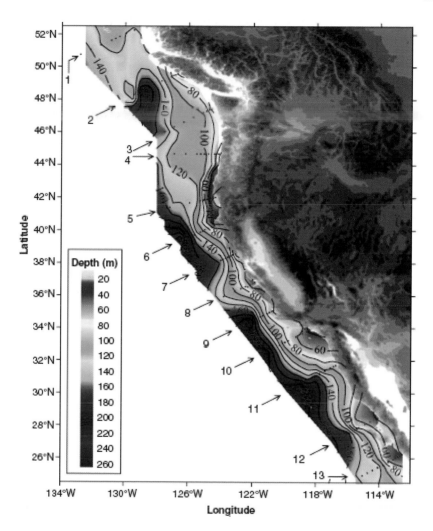

FIGURE 2.9 Distribution of the depths of the undersaturated water (aragonite saturation < 1.0; pH < 7.75) on the continental shelf of western North America from Queen Charlotte Sound, Canada, to San Gregorio Baja California Sur, Mexico. On transect line 5, the corrosive water reaches all the way to the surface in the inshore waters near the coast. The black dots represent station locations. (Feely et al., 2008)

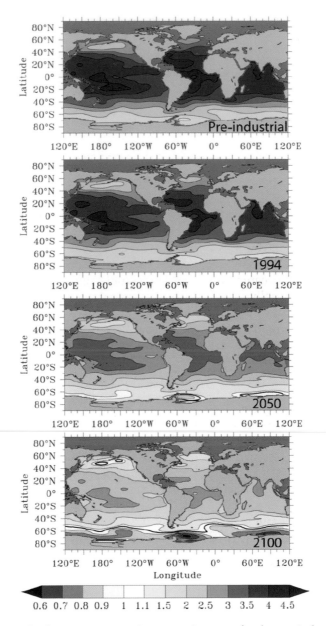

FIGURE 2.10 Surface water aragonite saturation state for the pre-industrial ocean (nominal year 1765), and years 1994, 2050, and 2100. Values for years 1765 and 1994 were computed from the global gridded data product GLODAP (Key et al., 2004), whereas the saturation state for years 2050 and 2100 are the median of 13 ocean general circulation models forced under the IPCC's IS92a "business-as-usual" CO_2 emission scenario (Orr et al., 2005). (Fabry et al., 2008b)

tion state of surface oceans under the "business-as-usual" (i.e., IPCC's IS92a) emissions scenario through the year 2100 (Orr et al., 2005; Fabry et al., 2008b). Under current rates of CO_2 emissions, models project that surface waters of the Southern Ocean, the Arctic Ocean, and parts of the subarctic Pacific will become undersaturated with respect to aragonite by the end of this century, and, in some regions, as early as 2023 (Orr et al., 2005; Steinacher et al., 2009).

2.3 CONTEXT AND CONSTRAINTS FROM THE GEOLOGIC PAST

Information about past changes could be helpful for understanding ongoing changes and their consequences. On time scales of thousands of years and longer, the pH of the ocean is determined primarily by the cycling of $CaCO_3$ (and some silicate) minerals which are dissolved on land, carried by rivers to the ocean where they are reprecipitated, and eventually buried in sediments. Ocean acidification results from the fact that this natural oceanic $CaCO_3$ cycle cannot keep up with the rapid rise in CO_2. But eventually, over thousands of years, changes in $CaCO_3$ cycling will neutralize most of the excess acidity and restore the pH of the ocean to near-present-day value. Natural glacial-interglacial changes in atmospheric CO_2 over the past 800,000 years, which are recorded in ice cores, occurred over thousands of years, thus reducing the magnitude of change in ocean pH for a given increase in atmospheric CO_2 and allowing time for the $CaCO_3$ cycle to keep up (Ridgwell and Zeebe, 2005).

In the deeper geologic past, millions of years ago, atmospheric CO_2 concentrations were much higher than today, giving the Earth a warm climate similar to the present-day tropics all the way to the high latitudes. This is often referred to as "hot house" conditions as compared to present-day "ice house" conditions. Again, in these hot-house cycles, because the CO_2 concentration changed over millions of years, the $CaCO_3$ cycle stabilized the pH of the ocean to these CO_2 changes, as evidence by massive $CaCO_3$ deposits from those periods. While glacial-interglacial cycles and hot house-ice house cycles provide information regarding the response of the ocean carbon cycle to changes in ocean pCO_2 over thousands and millions of years, they are not good analogs to current acidification of the ocean by anthropogenic CO_2.

2.4 MITIGATION AND GEOENGINEERING

There is currently a great deal of international interest in mitigating the impacts of climate change. However, this leads to the question of how these mitigation strategies will affect ocean acidification and how ocean acidification itself can be mitigated. Clearly, all mitigation strategies for climate change that reduce CO_2 inputs to the atmosphere will also reduce ocean acidification. These include increasing energy efficiency, shifting energy sources from fossil fuels to nuclear and renewables, and implementing carbon capture and storage technologies (Pacala and Socolow, 2004). Similarly beneficial would be carbon management approaches that remove CO_2 from the atmosphere through biological sequestration on land (e.g., afforestation, soil conservation) or industrial-scale geochemical approaches (Stephens and Keith, 2008). But geoengineering solutions designed to slow climate warming without reducing atmospheric CO_2 concentration, such as injection of sulfate aerosol precursors into the stratosphere (Crutzen, 2006), will not reduce ocean acidification (Wigley, 2006; Boyd, 2008). On a regional scale, in coastal and estuarine waters where acidification in surface waters may result partly from pollution such as acid rain or in bottom waters from eutrophication induced by excessive nutrient inputs, limiting emissions of air or water pollutants may be effective as a mitigation strategy.

Management strategies designed to sequester CO_2 in the ocean could potentially exacerbate ocean acidification in intermediate or deep waters. Iron fertilization of surface waters has been suggested as a potential approach for boosting primary production in regions that are iron-limited, thus increasing the export of organic carbon to the subsurface as discussed in the next chapter (Boyd et al., 2007). Critics of iron fertilization have questioned its efficiency at sequestering CO_2 and pointed out the difficulty in predicting its ecological consequences. Ocean acidification could affect the efficiency of iron fertilization and its potential consequences by modifying the biological availability of iron in surface seawater (see section 2.2). If effective, iron fertilization would increase the rate of penetration of CO_2 into intermediate waters, thus accelerating acidification in those water masses. A similar effect would result from direct injection of CO_2 into intermediate or deep ocean waters. Enhanced deep sea acidification could also occur as a result of leakage from sub-seabed CO_2 sequestration (Blackford et al., 2009) either in sediments (House et al., 2006) or bedrock (e.g., oil and gas fields, salt domes, etc.) (Caldeira and Wickett, 2005).

The effectiveness of direct ocean CO_2 injection techniques could theoretically be enhanced and the resulting acidification minimized by first reacting CO_2 with a base, neutralizing the carbonic acid, and producing primarily bicarbonate. Alternatively a base could be added directly to seawater. Most proposed schemes use carbonate rock (e.g., limestone) as the base and differ mostly in the techniques used to accelerate $CaCO_3$ dissolution (Rau and Caldeira, 1999; Caldeira and Rau, 2000; Golomb et al., 2007; Harvey, 2008; Rau, 2008). Sodium hydroxide (NaOH) could also be produced from water electrochemically with the co-produced hydrochloric acid (HCl) being neutralized by silicate rocks (House et al., 2007). Given that neutralization of CO_2 requires an equivalent amount of base (1:1 molar ratio), the logistics and resource demands for neutralizing a significant fraction of the 2 Pg C ($\sim 10^{14}$ moles CO_2) per year taken up by the ocean are likely to be prohibitive. But such mitigation strategies might be feasible on a local or regional scale.

3

Effects of Ocean Acidification on the Physiology of Marine Organisms

The dissolution of anthropogenic carbon dioxide (CO_2) into the ocean is causing a series of changes in ocean water chemistry: an increase in the CO_2 concentration, a decrease in the calcium carbonate saturation (Ω) and pH, and a change in the chemistry of many biologically important chemical species, as discussed in Chapter 2. These chemical changes will affect a range of biological processes in marine organisms, including the precipitation of calcium carbonate, fixation and respiration of CO_2, regulation of internal pH, and uptake of nutrients for growth. The questions are by what mechanisms will higher CO_2 affect an organism's physiology, to what degree will this affect the fitness of different organisms, and how will high CO_2 effects on individual organisms be dampened or amplified at the ecosystem level. This chapter reviews what is known of the effects of ocean acidification on a range of biological processes that have been studied in various organisms. It focuses on processes that are likely to be affected by acidification, both those that are common to many organisms (i.e., calcification and pH control) and those that affect primary production, which provides the principal influx of organic material and energy to marine ecosystems (i.e., photosynthetic carbon fixation, nutrient uptake, and nitrogen fixation). Chapter 4 reviews how these effects on individual organisms may scale up to the ecosystem level.

3.1 CALCIFICATION

Calcium carbonate ($CaCO_3$) is one of the most common building materials used in the formation of skeletons, shells, and other protective

45

structures in the marine biota. Marine calcifying organisms include many taxonomic groups and occupy diverse ecological niches. Important examples include photosynthetic primary producers (e.g., coccolithophores and coralline algae), zooplankton (e.g., pteropods), mollusks (e.g., clams, mussels, and oysters), crustaceans (e.g., crabs and lobsters), and animals that harbor photosynthetic symbionts (e.g., reef-building corals, some planktonic foraminfera). In most of these organisms, $CaCO_3$ is the principal constituent of the "hard part." But in some organisms, only a part of the exoskeleton is calcified (e.g., the calcite ossicles of sea stars), while, in others, calcium carbonate is integrated into an organic exoskeleton structure (e.g., lobster and crab shells). These $CaCO_3$ structures are most often in the form of calcite, aragonite, high-magnesium calcite, or a mixture of these mineral forms, and the mineral form may change through the development of the organism (Politi et al., 2004; see also Box 2.3).

Most calcifying organisms studied so far show a decrease in calcification or shell weight (either a slower rate of calcification or a decrease in the mass of $CaCO_3$ per individual) in response to elevated CO_2 and reduced pH. This is the best-documented and most widely observed biological effect of the acidification of seawater. It has been reported in a range of organisms, including coccolithophores, foraminifera, mussels, urchins, oysters and other bivalves, corals, and coralline algae (e.g., see Fabry et al. 2008b; Ries et al., 2009). In some organisms, a significant reduction in calcification was observed for a decrease in pH of 0.2-0.4 units, in the range predicted to occur over the next century; in others, a significant effect was only observed under more severe acidification. A few studies have shown that some calcifying organisms are insensitive to seawater acidification, or even increase calcification over the range of pH projected for the next century (Ries et al., 2009; Wood et al., 2008; Miller et al., 2009). In coccolithophores, the effect can be complicated by the increase in growth rate caused by high CO_2 (see below), such that the calcification rate per cell may increase while the ratio of inorganic to organic cellular carbon may decrease (Iglesias-Rodriguez et al., 2008). Note that increased calcification is not necessarily an indication of increased health of the organism, and there is some early evidence that other processes can be affected by the strain introduced by accommodation of the increased CO_2 (Wood et al., 2008). It is also possible that species that live in environments where pH and CO_2 concentrations are variable may be more tolerant of the overall increase in acidity, though this hypothesis has not been tested.

In some groups of organisms, a decrease in calcification is associated with more frequent malformations of the carbonate structures (e.g., coccolithophores; Riebesell et al., 2000; Langer et al., 2006), smaller and thinner shells in foraminifera (Moy et al., 2009) and mollusks (Miller et al., 2009; Talmage and Gobler, 2009), slower shell extension rates (e.g., mollusks;

Miller et al., 2009), and weakened shells (e.g., barnacles, mollusks) (Bibby et al., 2007; Clark et al., 2009; McDonald et al., 2009; Tunnicliffe et al., 2009). In reef-building corals, which have been studied most extensively, a wide range of responses has been observed, but on average, a doubling of preindustrial atmospheric CO_2 concentration resulted in about a 10–60% decrease in calcification rates (Langdon and Atkinson, 2005; see also Figure 3.1 as an example, section 4.1, and Appendix C). In some species, seawater acidification led to reduced rates of larval development and increased larval mortality (e.g., echinoderms) as a result of the instability of the nascent calcified structures which are often less well crystallized than the mature form. It must be noted, however, that the physiological role of calcification is not always clear. For example, laboratory cultures of coccolithophores that have lost the ability to calcify grow at normal rates (Rost and Riebesell, 2004). Some species of corals can grow well in cultures without precipitating aragonite, even though the very structure of a coral reef depends on the precipitation of the mineral (e.g., Fine and Tchernov, 2007).

The spontaneous precipitation of $CaCO_3$ in seawater requires a high degree of supersaturation of the mineral (i.e., $\Omega \gg 1$) (which is proportional to the carbonate ion $[CO_3^{2-}]$ concentration when pressure, temperature, and calcium ion concentration are kept constant). Within organisms, this is achieved by controlling Ω at the site of calcification at a value generally higher than that of seawater (e.g., Al-Horani et al., 2003; Furla et al., 1998; Bentova et al., 2009) through a process that involves pumping of various ions into specialized cellular compartments. Despite the fact that organisms control internal Ω with these processes, calcification has been observed to correlate well with the external value of Ω in many experiments and several taxa (see Figure 3.1); therefore, the external Ω and CO_3^{2-} concentration may serve as indicators of the calcification response caused by acidification. There are several hypotheses regarding the correlation between external Ω and biological calcification; for example, acidification of the external medium may increase the energetic cost of calcification in some organisms. The energetic cost of calcification should depend on the underlying biochemical mechanisms, which are presently not well understood and are likely to differ widely among taxonomic groups. Marubini and others (2008) addressed multiple hypotheses to explain why acidification causes a decrease in coral calcification rates and suggested that decreases in intracellular or extracellular pH, or shifts in the buffering capacity of the calcifying fluid were likely. A recent study on a temperate coral provides evidence that the calcification response reflects changes in the proton pumping capacity, which is necessary to maintain the high saturation states of the internal calcifying fluid (Cohen et al., 2009). This is supported by results from an analytical survey of calcification responses across multiple taxa, which suggests that the calcification response correlates with an organism's ability to regulate its internal pH, as well as other factors such as the degree of shell protection by organic coatings, shell mineralogy, and whether the organism utilizes photosynthesis (Ries et al., 2009).

As ocean acidification decreases the CO_3^{2-} concentration, it decreases the degree of supersaturation of $CaCO_3$ in the upper water column, brings the saturation horizons closer to the surface, and may result in some organisms being exposed to undersaturated (corrosive) waters. This may result in the dissolution of previously precipitated minerals at shallower depths. For example, when exposed to the level of aragonite undersaturation expected to occur at high latitudes by the year 2100 (see Chapter 2), shell dissolution was visually evident in a subarctic pteropod species within 48 hours (Orr et al., 2005). Such conditions of undersaturation will happen sooner for the more soluble aragonite, which is formed by corals, some calcifying macroalgae, and some mollusks, than for calcite, produced by coccolithophores, foraminifera, echinoderms, many deep-sea corals, some mollusks, and crustaceans (Pearse et al., 1987). In some species, the calcareous shells and skeletons are protected from dissolution by organic coatings, but in others they are largely exposed to the surrounding seawater (e.g., some bivalve mollusks such as scallops and oysters (Ries et al., 2009). There are apparently large differences among taxa, with some organisms being able to maintain calcified structures in highly corrosive waters (Wood et al., 2008; Tunnicliffe, 2009).

Overall, the acidification of seawater should prove unfavorable for most calcifying organisms, and this is likely to constitute a major negative effect on the marine biota. But it must be emphasized that despite exten-

FIGURE 3.1 *[facing page]* Examples of the effect of decreasing CO_3^{2-} concentration (A) or of increasing pCO_2 (B and C) on calcification rates in various taxa. As pCO_2 increases, the carbonate ion concentration in the water will decrease and hence the aragonite saturation will also decrease. (A) In the Biosphere 2 coral mesocosm, the system was perturbed by adjusting the carbonate ion concentration (Langdon et al., 2000). In (B), the blue mussel *Mytilus edulis* and Pacific oyster *Crassostrea gigas* (Gazeau et al., 2007), and (C), coccolithophorids *Emiliania huxleyi* and *Gephyrocapsa oceanica* (Riebesell et al., 2000), pCO_2 was increased to mimic the effect of higher atmospheric pCO_2 on the carbonate saturation state. For cross-comparison, dashed lines have been added to these plots to indicate pCO_2 concentrations of 280 ppm (preindustrial concentration), 390 ppm (current-day concentration), 560 ppm (twice preindustrial concentration), and 780 ppm (estimated concentration in the year 2100). In A, the pCO_2 concentrations are roughly estimated based upon average aragonite saturation level (Ω_{arag}) across the coral reefs.

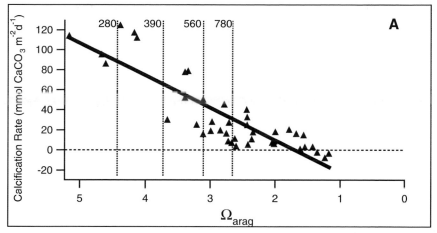

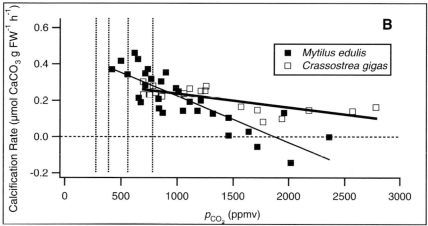

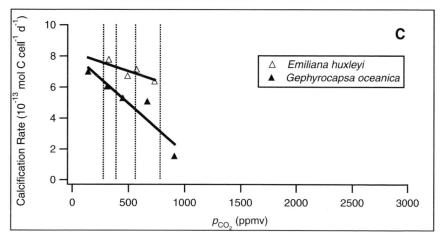

sive research efforts we still have a poor understanding of the mechanisms and regulation of the calcification process in marine organisms.

3.2 INTERNAL PH CONTROL AND OTHER METABOLIC PROCESSES

Biological membranes are generally highly permeable to dissolved CO_2, therefore, dissolved CO_2 will equilibrate across membranes following the concentration gradient. Dissolved CO_2 in internal fluids tends to form bicarbonate and free hydrogen ions, acidifying the medium as it does in seawater. Most heterotrophic organisms excrete CO_2, produced as a by-product of metabolic activity, by utilizing a concentration gradient from high internal to the lower, external dissolved CO_2. If external dissolved CO_2 rises, the efficiency of this mechanism will decrease, potentially affecting acid-base balance in the organism.

Most heterotrophic organisms maintain internal pH lower than normal seawater—(Hochachka and Somero, 2002). Bacteria often have optimal intracellular pH values between 7.4 and 7.8 that they maintain over a fairly wide range in external pH (Booth, 1985; Padan et al., 2005). The internal pH of multicellular marine organisms is also typically lower than seawater, with a progressive decrease in pH from external to internal spaces: extracellular fluids (blood spaces, fluids surrounding cells) have a pH lower than external seawater, and intracellular pH is lower (~0.4 pH units) than that of the extracellular fluids. Intracellular pH is tightly modulated because many metabolic processes are regulated by small shifts in the pH of the medium or depend on a small proton gradient across membranes. Hence, the metabolism of the organism is usually linked to the homeostasis of internal pH as well as the internal to external pH gradient (Pörtner et al., 2004). As a consequence, an increase in the environmental CO_2 concentration from ocean acidification could perturb the internal acid-base balance of organisms, potentially affecting a variety of cellular functions ranging from protein synthesis to calcification.

The ability to buffer or control internal pH varies considerably among organisms, in part related to their complexity. In all organisms, partial pH control is achieved through the passive buffering capacity of the internal fluids and the active regulation of various ion pumps (Seibel and Walsh, 2003). Multicellular organisms typically have greater passive buffering capacities, and many can control the pH of their body fluids by secreting or eliminating acid or base through specialized organs (Melzner et al., 2009). These homeostatic mechanisms allow some aquatic organisms to acclimate to a range of external pH/pCO_2. But the metabolic cost of this acclimation may slow the growth or decrease the fitness of some organisms and some may not be able to acclimate at all (e.g., Wood et al., 2008).

Many experiments on metabolic costs have been conducted at lower pH and higher CO_2 concentrations than expected from ocean acidification over the next century. Nonetheless, this work provides a mechanistic understanding of how animals respond to internal acidosis caused by high environmental CO_2 levels. The effects on various metabolic functions may lead to metabolic depression, which decreases all aerobic activities of the organism (Pörtner et al., 2004).

In taxa with respiratory proteins (e.g., hemoglobin, hemocyanin), extracellular acidosis affects oxygen transport and respiratory efficiency due to the reduced oxygen affinity of these proteins at lower pH (Seibel and Walsh, 2003). Animals that actively regulate internal pH will have a greater metabolic demand to meet the high energetic cost of pumping ions across membranes, but the decreased affinity of the respiratory proteins will make it more difficult for the organism to meet that metabolic demand due to the reduction in overall aerobic respiration (Pörtner et al., 2000).

Tolerance for acidification varies greatly among phyla and is linked to metabolic rate and, in turn, to the transport capacities for oxygen and CO_2. Because metabolic activity generates by-products that lower pH and increase CO_2 in animal tissues, many highly active animals have mechanisms that help them regulate pH and CO_2 levels in their internal fluids and tissues. Those animal groups (e.g., mammals, fishes, and some mollusks), have a high capacity for oxygen and CO_2 transport and exchange and appear to be tolerant of more acidic environmental conditions, at least over short periods that are similar to the conditions resulting naturally from bouts of high activity (Melzner et al., 2009). In contrast, many marine invertebrate taxa that have been examined have less developed gas exchange and acid-base regulatory capacities, and are expected to have lower tolerance to acid-base disruption caused by ocean acidification (Melzner et al., 2009). Still, there is quite a lot of variation across taxa and little is known about the extent of this variation since some groups, such as gelatinous zooplankton, have not yet been studied. Because of the high energetic demands of acid-base regulation, the ability of organisms to cope with acid-base disturbance also varies among habitats, with those inhabiting energy-poor habitats (e.g., deep-sea environments) exhibiting less tolerance than others.

There are likely to be many other important effects of acidification beyond internal pH control, particularly in higher organisms such as finfish. For example, a recent study showed impaired olfactory discrimination and homing ability in the larvae of the orange clownfish *Amphiprion percula* (Munday et al., 2009). Currently, these effects are almost completely unknown. To date, the state of knowledge concerning the effects of decreasing pH and increasing CO_2 on most marine organisms is sparse. Although many of the underlying physiological mechanisms are understood in some detail, knowledge of the metabolic consequences for individual performance remains weak. Understanding is particularly poor concerning the sensitivities of various life stages of marine organisms, although initial studies suggest vulnerability of early life history phases of several groups such as bivalves and some echinoderms (Talmage and Gobler, 2009; Dupont and Thorndyke, 2009). Even less is known about the cumulative, lifelong effects of a lower pH environment in terms of how it will affect the performance, growth, survival, and fitness of individuals, especially when combined with other likely stressors.

3.3 PHOTOSYNTHETIC CARBON FIXATION

In the oceans, photosynthesis—the formation of organic matter using sunlight energy—is carried out chiefly by microscopic phytoplankton and, to a lesser extent, by macroalgae and seagrasses. For this purpose, photosynthetic organisms must acquire, among other things, inorganic carbon (i.e., CO_2) from seawater. Dissolved CO_2 is the substrate used in the "carbon fixation" step of photosynthesis, not the more abundant forms of dissolved inorganic carbon. Thus CO_2, which is at low concentration in the ambient water, must be concentrated at the site of fixation; this is a difficult and energy-consuming process because the CO_2 molecule diffuses readily through biological membranes and continuously leaks out of cells. It is thus expected that an increase in the CO_2 concentration of surface seawater would facilitate marine photosynthesis and lead in some cases to an increase in primary production (i.e., the rate of organic matter synthesis per unit time and unit area of the ocean).

Enhancement of photosynthesis under high-CO_2 conditions has indeed been observed in a number of experiments with some, but not all, species of marine algae. For example, an enhancement of photosynthesis at high CO_2 has been seen in calcifying coccolithophore species that form massive blooms in many oceanic regions (Riebesell et al., 2000; Zondervan et al., 2002), but not in some marine diatoms (Burkhardt et al., 1999) or in the symbiotic dinoflagellates of corals (zooxanthellae) (Schneider and Erez, 2006). In another example, two types of cyanobacteria, *Synechococcus* and *Prochlorococcus*—representing the two dominant open ocean cyanobacteria species—responded differently to high CO_2 conditions (Figure 3.2). In a few cases a decrease in photosynthesis has been seen at elevated CO_2 (Reynaud et al., 2003). Increased photosynthetic rate does not always translate to higher growth; it appears that at high CO_2, some phytoplankton species release a sizeable fraction of their photosynthate as extracellular organic matter (Engel, 2002).

The effect of increasing CO_2 concentration on photosynthesis depends on the underlying biochemical mechanisms involved in concentrating CO_2

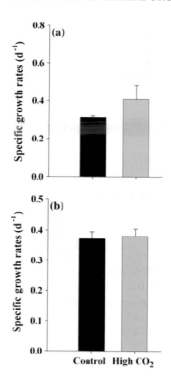

FIGURE 3.2 The effect of CO_2 on the growth of (a) *Synechococcus*, which shows an increase in growth rate in high CO_2 and (b) *Prochloroccus*, which shows no effect of high CO_2 on growth rate. (Fu et al., 2007)

at the site of fixation. These mechanisms have not all been elucidated, but are known to differ between different photosynthetic marine organisms. For example, some diatoms concentrate CO_2 using a mechanism similar to the one that operates in so-called "C_4 plants" such as sugar cane and maize (Reinfelder et al., 2000, 2004; McGinn and Morel, 2008). The difference in the CO_2 concentrating mechanisms presumably explains the different responses of phytoplankton species.

Overall, an increase in the CO_2 concentration is expected to enhance rather than decrease the growth of photosynthetic organisms and the production of organic matter in the ocean. But this effect is generally modest and appears variable among species; it may thus lead to a shift of dominant species of phytoplankton (see also Chapter 4.2). In most cases, the potential enhancement of primary production by CO_2 will be constrained by nutrient limitation. These projections are based on limited data in the marine environment, but they are supported by the analogy with land plants, which possess similar underlying photosynthetic mechanisms. A large number of observations on terrestrial plants exposed to high CO_2 show a boost in photosynthesis and a differential response among species.

3.4 NUTRIENT ACQUISITION AND LIMITATION

In different oceanic regions, primary production by phytoplankton can be limited by the availability of various key nutrients, most commonly nitrogen, phosphorus, or iron. Ocean acidification may alter the availability of nutrients in three ways: (1) by changing the chemical forms of nutrients in the water; (2) by changing the activity of enzymes that convert nutrients into useable forms; and (3) by changing the nutrient requirements of the phytoplankton.

The acquisition of nutrients depends on their chemical form—the chemical "species"—present in the water. This is particularly true for trace metals such as zinc, cobalt, nickel, or iron, which are essential for various biochemical processes inside cells. These metals are readily taken up when present as free ions or ions bound to chloride, hydroxide, or other inorganic species, but require specialized uptake machinery when bound in organic complexes (Morel et al., 2003). Because the bulk of most bioactive metals are bound in organic complexes and the extent of such binding is generally sensitive to pH, it is thought that metal bioavailability might be affected by the acidification of surface seawater. The most important case is that of iron, which limits phytoplankton growth in large parts of the equatorial Pacific and high latitude oceanic regions. An increase in organic complexation makes dissolved iron less bioavailable as pH decreases (Shi et al., 2010); however, this effect may be offset by other effects of pH on the cycle of iron in surface seawater including an increase in the solubility of iron oxides and an enhancement in the light-induced redox cycle of iron.

The bioavailability of nutrients may also be affected through the influence of pH on biochemical rather than chemical processes. In some cases, phytoplankton can use enzymes to convert nutrients that are not readily available into a useable form. For example, free phosphate in seawater can be readily taken up by phytoplankton but, when its concentration is very low, some organisms can use phosphate bound in organic compounds. In this case the phosphate must first be cleaved enzymatically from the organic molecule before being utilized. In the range of pH relevant to the surface ocean at present and in the future, the activity of the enzyme responsible for this cleavage (known as alkaline phosphatase) decreases rapidly with decreasing pH (Figure 3.3). Since the enzyme operates outside the cell, it responds directly to acidification of the external medium. Therefore, organisms that depend on organic phosphate for growth will have a more difficult time acquiring phosphate in an acidified ocean, which may negatively affect their growth. Similar changes in bioavailability may occur for some organic forms of nitrogen such as amino acids or amines, which are also acquired by some phytoplankton species through extracellular enzymatic processes (Palenik and Morel, 1990, 1991a, 1991b).

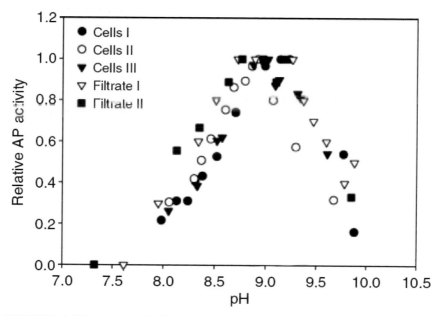

FIGURE 3.3 The activity of alkaline phosphatase (AP) vs. pH in *Emiliania huxleyi* cultures; three measurements were taken for AP on the cell surface, while two were taken in the culture filtrate. Note that AP activity decreases as pH decreases (in the range expected from acidification). (Xu et al., 2006)

Primary production—the amount of organic matter that is synthesized per area of surface seawater per unit time—generally depends on the rate of supply of the limiting nutrient. A change in primary production requires either a change in nutrient requirement (e.g., an increase in unit of carbon fixed per unit of the limiting nutrient) or in the supply of the limiting nutrient. There is evidence that at high CO_2, phytoplankton produce organic matter with a different elemental composition, particularly a higher carbon to nitrogen (C:N) ratio, suggesting that the phytoplankton are able to change their N requirement (e.g., Riebesell et al., 2007; Bellerby et al., 2007; Fu et al., 2007; Hutchins et al., 2009). Such an effect would lead to an increase in the quantity of organic carbon formed per unit of limiting nutrient. It must be noted that a reduction in the nutrient supply may result from the increased stratification of surface seawater caused by increases in global temperatures. In this way the effects of increasing CO_2 on climate and on ocean chemistry may compound or partly alleviate each other (see Chapter 2).

3.5 NITROGEN FIXATION

Nitrogen fixation is the process of converting atmospheric nitrogen gas (N_2), which cannot be used by organisms as a source of nitrogen for biosynthesis, into ammonium (NH_4^+), a form readily available to the biota. In the oceans, this process is predominantly carried out by a few specialized cyanobacteria. Nitrogen fixation represents a major input of "new" nitrogen to marine ecosystems and is thus a key in controlling primary production in large regions of the world's oceans. Nitrogen fixation is an "expensive" biochemical process that requires synthesis of a complex, iron-rich enzyme and uses large amounts of energy. Changes in the availability of iron and, possibly, other nutrients may thus change the rate of N_2 fixation in the oceans. In one study, elevated CO_2 decreased growth and N_2 fixation rates in the heterocystous cyanobacterium *Nodularia spumigena* (Czerny et al., 2009). But a few experiments with cultures of the dominant marine N_2 fixer *Trichodesmium* have shown a substantial increase in the rate of N_2 fixation at elevated CO_2 concentrations (Hutchins et al., 2007; Levitan et al., 2007; Barcelos e Ramos et al., 2007; Kranz et al., 2009). Some recent experiments with natural populations of *Trichodesmium* incubated at high CO_2 appear to confirm the laboratory results (Hutchins et al., 2009; see also Chapter 4.2).

3.6 ACCLIMATION AND ADAPTATION

Acclimation is the process by which an organism adjusts to an environmental change that gives individuals the ability to tolerate some range of environmental variability. Although acclimation may allow individual organisms to survive a certain amount of stress, metabolic performance, including growth and reproduction, may be depressed in scale with the magnitude of the environmental perturbation. Therefore, acclimation may have population-level effects even though survival is increased at the level of the individual. The potential for individuals of most species to acclimate to higher CO_2 and lower pH is not known, but will become increasingly important as ocean CO_2 levels rise.

Adaptation is the ability of a population to evolve over successive generations to become better suited to its habitat. Adaptation to changing ocean chemistry is likely on some level for most taxa that have sufficient genetic diversity to express a range of tolerance for ocean acidification. Rates of adaptation are linked strongly to generation times, which range from days for many microbial and unicellular organisms (e.g., allowing for >30,000 generations by 2100) to as long as decades (e.g., allowing for ~10 generations by 2100) for some slow-growing, long-lived marine animals. It remains unknown whether populations of most species possess both the genetic diversity and a sufficient population turnover rate to allow adaptation at the expected rate and magnitude of future pH/pCO_2 changes. It is conceivable that some reef calcifiers or cold water corals could adapt to ocean acidification if they evolve a calcification mechanism that allows them to precipitate $CaCO_3$ at normal rates, but this type of adaptation has not been documented in corals. Survival of these organisms depends on their capacity to cope with skeletal loss by a change in life history (i.e., shift to cryptic existence), defenses (e.g., toxin production), or other means. Adaptation to compensate for weaker or smaller skeletons has not been demonstrated, but this topic has barely been investigated (e.g., Bibby et al., 2007).

The persistence of various taxa under increasing ocean acidification will depend on either the capacity for acclimation (plasticity in phenotype within a generation) or adaptation (plasticity in genotype over successive generations) or a combination of both. The relative capabilities of various taxa in terms of both acclimation and adaptation will likely influence the composition of marine communities and therefore result in a range of consequences for marine ecosystems.

4

Effects of Ocean Acidification on Marine Ecosystems

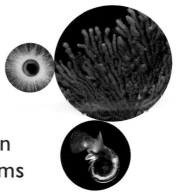

Ecosystems are defined by a complex suite of interactions among organisms and also between organisms and their physical environment; a disturbance to any part may lead to cascading effects throughout the system. Ocean acidification has the potential to disturb marine ecosystems through a variety of pathways. Differential sensitivities will result in ecological winners and losers, as well as temporal and spatial shifts in interactions between species (e.g., shifts in the timing of zooplankton development relative to food availability; Pörtner and Farell, 2008), leading to changes in predator-prey, competitive, and other food web interactions. There may also be changes in habitat quality and effects on other ecological processes such as nutrient cycling. Many of the physiological changes from ocean acidification are expected to affect key functional groups –species or groups of organisms that play a disproportionately important role in ecosystems. These include expected effects on phytoplankton, which serve as the base of marine food webs, and on ecosystem engineers, which create or modify habitat (e.g., corals, oysters, and seagrasses). Such changes may lead to wholesale shifts in the composition, structure, and function of these systems and ultimately affect the goods and services provided to society (see Chapter 5). While it is important to understand how ocean acidification will change ocean chemistry and the physiology of marine organisms, as reviewed in chapters 2 and 3, what is equally critical is to understand how these effects may scale up to populations, communities, and entire marine ecosystems. Such changes are likely to be difficult to predict, particularly where more than one species or functional group will be affected by ocean acidification. In general, higher trophic levels, including most finfish, will likely be sensitive to ocean acidification through changes in the quantity or composition of the food available, although there may be direct physiological effects on some fish species at high pCO_2 (see Chapter 3). The difficulty in predicting ecosystem change is compounded by other simultaneous stressors occurring in the oceans now (e.g., pollution, overfishing, and nutrient eutrophication) and in association with climate change. For example, it is projected that surface waters will become warmer, the upper water column will become more stratified, and the supply of nutrients from deep waters and from the atmosphere will change as a result of climate change. Whether these changes, in combination with the effects of ocean acidification, will have synergistic, antagonistic, or additive effects is unknown, but multiple stressors are likely to affect marine ecosystems at multiple scales.

Several previous reports have identified marine ecosystems that are most likely to be at risk from ocean acidification (e.g., Raven et al., 2005; Fabry et al., 2008b). This chapter begins by describing what is known and not known about ecosystem effects of ocean acidification for five vulnerable ecosystems: tropical coral reef, open ocean plankton, coastal, deep sea, and high latitude ecosystems. This is not an exhaustive review of all possible ecological effects, but is instead an overview of the ecosystems that have been identified as most vulnerable to acidification. The chapter looks at examples of high-CO_2 periods in the geologic past for possible information on the ecological response to current acidification. It also examines general principles regarding biodiversity, possible thresholds in ecological systems, and managing ecosystems for change.

4.1 TROPICAL CORAL REEFS

Some of the most convincing evidence that ocean acidification will affect marine ecosystems comes from warm water coral reefs. Coral reef ecosystems are defined by the large, wave-resistant calcium carbonate structures, or reefs, that are built by reef calcifiers. The structures they build provide food and shelter for a wide variety of marine organisms (Figure 4.1). There are hundreds of reef-building species; the predominant calcifiers on coral reefs are zooxanthellate corals, which produce hard aragonite skeletons, and calcifying macroalgae,[1] which produce high-Mg calcite and aragonite. These groups produce the bulk of the calcium carbonate that make up the reef structures, which in turn support the high biodiversity of coral reef ecosystems. Recent analyses illustrate that

---

[1] There are two types of calcifying macroalgae that are important to reef formation in tropical coral reef ecosystems: crustose coralline red algae (coralline algae) from the family Corallinaceae and calcifying green algae (genus *Halimeda*)

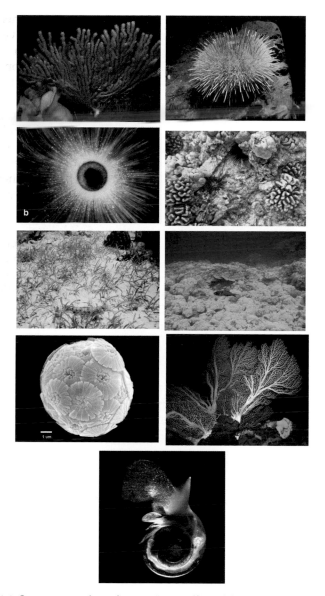

FIGURE 4.1 Some examples of organisms affected by ocean acidification. Red coral (photo courtesy of Jim Barry, MBARI); Sea urchin (photo courtesy of Jim Barry, MBARI); Foramaniferan (photo courtesy of Howard Spero, University of California, Davis); Coral and sea urchins (photo courtesy of Susan Roberts, NRC); Sea grass (photo courtesy of Richard Zimmerman, Old Dominion University); Tropical coral reef and fish (photo courtesy of Susan Roberts, NRC); Coccolithophores (photo courtesy of Mitch Covington, BugWare Inc.); Deep-sea Gorgonian bubblegum coral (photo courtesy of MBARI); and Pteropod (photo courtesy of Russ Hopkroft, University of Alaska, Fairbanks).

reef ecosystems have served as "cradles of evolution" throughout Earth's biological history (Kiessling et al., 2010); that is, more marine species have originated in reef ecosystems than in any other. As a consequence, a decrease in the resilience of coral reefs or loss of coral reef habitat may adversely affect marine biodiversity in the short and long term. These ecosystems also provide a variety of services to humans, including recreation, fisheries, and coastal protection.

Ocean acidification poses a variety of risks to coral reef ecosystems. A critical vulnerability is the potential for ocean acidification to affect the reef structure itself. Acidification may decrease reef growth by reducing calcification rates, reproduction, and recruitment. It may also increase the dissolution or erosion of existing reef structures. Finally, acidification may indirectly result in the mortality of reef-builders.

The most obvious and best documented effect of ocean acidification is the depression of calcification rates, which will affect skeletal growth of the reef-building organisms. Decreased coral calcification rates are evident on the Great Barrier Reef, where records from massive corals show that calcification rates decreased by about 14% between 1990 and 2005 (De'ath et al., 2009), although the relative roles of increased temperature and ocean acidification could not be determined. Decreased skeletal growth in tropical reef-building corals and coralline algae has been well documented in high CO_2 conditions that result in ocean acidification (see Appendix C for a summary; see also reviews in Doney et al., 2009; Kleypas et al., 2006; Langdon and Atkinson, 2005). In stony corals, most studies indicate a 10-60% reduction in calcification rate for a doubling of preindustrial atmospheric CO_2 concentration. Differences among studies may reflect different species or experimental setups. Calcification rates in stony corals are affected by factors other than seawater carbonate chemistry, including light, nutrients, and particularly temperature. For example, studies on the effects of temperature show that calcification rates in corals peak near some optimal temperature (usually near the average summertime maximum), then decline at higher values (Clausen and Roth, 1975; Jokiel and Coles, 1977). As a result, increasing temperature from global climate change may initially offset the negative effect of acidification on calcification, but will eventually (and in some cases may already) work synergistically with acidification to decrease calcification. Calcification rates in tropical calcifying macroalgae may decrease even more strongly due to increasing CO_2. Several laboratory studies indicate that reef-building crustose coralline algae will calcify more slowly (e.g., 50% reduction; Reynaud et al., 2003; Anthony et al., 2008). Field studies seem to agree with these findings. In one study, coralline algae showed a higher calcification rate that correlated with the natural pH change from the photosynthetic drawdown of CO_2 when the algae grew in proximity to seagrasses (Semesi et al., 2009b). By comparison, in a study of a temperate benthic community, the abundance of crustose coralline algae decreased rapidly with proximity to a shallow submarine CO_2 vent, suggesting that coralline algae in this system could not survive at low pH (< 7.7) (Hall-Spencer et al., 2008; Martin et al., 2008). Similar to tropical reef corals, calcification rates of reef-building crustose coralline algae are affected more strongly by ocean acidification at elevated temperature (Anthony et al., 2008). There is little evidence that reef-building corals can adapt to decreased calcification under future ocean conditions.

Growth of reef structures relies not only on the calcification of adult corals, but also on successful recruitment of reef organisms, which is determined by gamete production, fertilization rates, larval development and settlement, and post-settlement growth. Theoretically, acidification could affect recruitment success but there is limited evidence of this and no consistent trends. In one study, ocean acidification did not affect either gamete production in one coral species or larval recruitment in another species (Jokiel et al., 2008). Another study also showed no effect on larval settlement, but did show significant decrease in post-settlement growth (> 50%; Albright et al., 2008). In general, there are few data on any of these aspects for reef-building species, making extrapolation to ecosystem effects difficult. Recruitment success may also be decreased through indirect effects on substrate. The presence of microbial biofilms or crustose coralline algae is important in coral recruitment success (Heyward and Negri, 1999; Negri et al., 2001; Webster et al., 2004; Williams et al., 2008). Reduction in the surface cover of newly recruited reef-building crustose coralline algae under future CO_2 conditions (Kuffner et al., 2008) could therefore affect recruitment of coral larvae.

While ocean acidification does not appear to cause direct mortality in corals, several studies suggest that the survival of both major calcifying groups will be indirectly affected by ocean acidification, mainly because of its effects on skeletal growth. Several reviews (Kleypas et al., 2006; Kleypas and Langdon, 2006) list multiple ways that reduced skeletal growth may impact coral survival rates, including the ability to withstand hydrodynamic and erosional forces, age of sexual maturity, rate of fragmentation, skeletal light-gathering properties (Enriquez, 2004), and recruitment success. In addition, there is some evidence that ocean acidification has contributed to bleaching, which can ultimately lead to coral mortality (Anthony et al., 2008).[2] Competition for space may also

---

[2] Most reef-building zooxanthellate coral species depend on photosynthetic endosymbionts—zooxanthallae—to provide energy. Bleaching refers to the loss of these zooxanthallae due to stress, resulting in a loss of color. While corals can regain their endosymbionts and recover from bleaching events, extended bleaching can also result in coral death (Glynn, 1996).

lead to loss of corals as they become more vulnerable to displacement by other organisms, including those that may benefit from ocean acidification, such as non-calcifying macroalgae. Macroalgae compete with corals by taking up suitable surface area, blocking sunlight, and through the sweeping action of algae in waves and currents that can abrade corals or prevent larval settlement on hard substrates. Conditions that favor macroalgal growth (e.g., high nutrients, elimination of herbivores) and/or slow coral growth (e.g., bleaching, disease, ocean acidification) lower the resilience of coral-dominated systems to disturbance and thus increase the likelihood of a regime shift. The density of several invasive macroalgae increased near natural CO_2 vents in the Mediterranean (Hall-Spencer et al., 2008), but little is known about the response of this or other groups that compete directly with corals for space. In some cases, an increase in non-calcifying primary producers on reefs (seagrasses and macroalgae) may counter the effects of ocean acidification, by drawing down CO_2 directly from the water column during photosynthesis (Palacios and Zimmerman, 2007; Semesi et al., 2009a). While many of these hypothesized effects seem logical, most have not yet been explicitly tested.

The overall calcium carbonate budget and reef-building capacity of a reef depend not only on carbonate production rates, but also on dissolution rates and carbonate removal rates due to erosion and sediment transport. Acidification has been shown to increase dissolution rates of coral reefs; in one extreme example, the skeletons of corals placed in seawater with pH of 7.3–7.6 dissolved completely (Fine and Tchernov, 2007). The combination of decreased calcification rates with increased dissolution rates will shift coral reefs from net production/accretion to net dissolution/erosion at some CO_2 threshold (Leclercq et al., 2000; Andersson et al., 2007; Yates and Halley, 2006; Silverman et al., 2009). Several studies indicate that crustose coralline algae will experience accelerated dissolution rates as ocean acidification proceeds and will experience net dissolution as pCO_2 levels approach 700 ppm, expected by the end of the century (Jokiel et al., 2008; Kuffner et al., 2008; Martin and Gattuso, 2009). This directly threatens the existence of this key functional group on coral reefs and in coralline algal-based ecosystems. One projection of reef building estimates that, due to reduced coral cover from bleaching and due to ocean acidification, all coral reefs will be in a state of net dissolution once atmospheric CO_2 concentration reaches 560 ppm (Silverman et al., 2009). The rapid loss of reef structure in the Galápagos following a severe bleaching event provides some evidence for this; the erosion rates of the Galápagos reefs were the highest recorded on any reef, which appears to be due in part to the naturally high CO_2 waters (400-700 ppm) in this region (Manzello et al., 2008).

The combination of potential effects of acidification on the ecosystem engineers of coral reefs—decreased calcification, increased dissolution, changes in recruitment and survivorship—will ultimately lead to changes in the reef structure. The function of calcium carbonate in reef ecosystems is widely recognized as important, but few studies have addressed what will happen as reef-building slows down. The dramatic loss of coral cover on many reefs has already resulted in "reef flattening" (a reduction in architectural complexity) that reduces the diversity of habitats and thus lowers the ability of the reef to support biodiversity (Alvarez-Filip et al., 2009). Ocean acidification is likely to exacerbate reef flattening. Loss of architectural complexity on reefs has been associated with changes in fish communities (Gratwicke and Speight, 2005; Pratchett et al., 2008), including the overall decline on Caribbean reefs (Paddack et al., 2009). Densities of important commercial species such as lobster have been linked to habitat complexity (Wynne and Côté, 2007), as well as recruitment of larval fish (Feary et al., 2007; Graham et al., 2007). Loss of structural complexity may also affect the recruitment of corals and other invertebrates, but this has not been examined. Finally, if reef structures suffer net erosion, then they lose their breakwater role, leaving coastlines and quiet-water habitats like mangroves more exposed to storm waves. The projected changes on reef structure are thus likely to have major consequences throughout tropical coral reef ecosystems.

4.2 OPEN OCEAN PLANKTONIC ECOSYSTEMS

The open ocean is not a uniform ecosystem; the components vary greatly by location. In open ocean systems, microscopic photosynthetic organisms—phytoplankton—which grow in the sunlit surface waters, serve as the base of diverse and complex food webs including zooplankton and larger free-swimming animals such as fish and marine mammals. Phytoplankton and bacteria also play an important role in cycling nutrients in open ocean ecosystems. Ocean acidification has been found to affect several key processes in open ocean planktonic ecosystems, including calcification, photosynthesis, and nitrogen-fixation. These changes affect the community composition of phytoplankton and zooplankton at the base of open ocean pelagic food webs; effects on these key functional groups may have cascading effects throughout the ecosystem. There may also be changes to the cycles of organic and inorganic carbon, oxygen, nutrients, and trace elements in the sea. In addition, the exchange of carbon dioxide and other climatically relevant trace gas species with the atmosphere may be modified, thus inducing feedbacks on the climate system.

The effect of acidification on calcification rates has been a major area of study because a number of the phytoplankton and zooplankton near the base of the food chain are calcifiers. Of the three major groups of planktonic calcifiers—coccolithophores, foraminifera, and pteropods (a planktonic snail) (Figure 4.1)—coccolithophores have been studied most widely. While experiments using monospecific cultures of coccolithophores revealed considerable species- and strain-specific differences in CO_2 responses (Rost et al., 2008; Langer et al., 2009), a consistent trend of decreasing calcification with increasing CO_2 has been seen in shipboard and mesocosm studies using mixed assemblages (Ridgwell et al., 2009). Studies on planktonic foraminifera and pteropods also indicate reduced calcification and increased calcium carbonate dissolution at elevated CO_2 (see Fabry et al., 2008b for review; Moy et al., 2009; see also section 4.5). It is presently unknown to what extent these responses affect the competitive abilities, susceptibility to viral attack, predator-prey interactions, or the fitness of calcifying plankton.

Reduced rates of calcification, along with the shoaling of the saturation horizons for calcium carbonate minerals to shallower depths will also affect the marine calcium carbonate cycle (see Chapter 2) through decreased $CaCO_3$ burial in sediments, additional carbon storage from increased production of extracellular organic carbon by phytoplankton (see below), and by the accelerated bacterial decomposition of organic matter at higher temperature. Ocean acidification can also affect processes related to photosynthetic activity, including increased rates of phytoplankton growth, primary production, and release of extracellular organic matter, as well as shifts in cellular carbon to nitrogen to phosphorus (C:N:P) ratios (e.g., Riebesell et al., 2007; Bellerby et al., 2007; Fu et al., 2007; Hutchins et al., 2009; see also Chapter 3). A shift in the ratio towards higher C:N and C:P at elevated pCO_2 was observed during a mesocosm study with a natural plankton community (Riebesell et al., 2007). Changes in the C:N and C:P ratios alter the nutritional value of phytoplankton and may adversely affect growth and reproduction of their consumers (e.g., as seen in copepods and daphnids; Sterner and Elser, 2002). A change in the composition of the biomass is one of the few mechanisms by which biology can alter ocean carbon storage (Boyd and Doney, 2003; Riebesell et al., 2009). If phytoplankton growing at high CO_2 produce and export biomass with a higher C:N ratio, it would make the ocean biological pump more efficient in exporting carbon to depth. In a mesocosm experiment, the net effect of this phenomenon was estimated to increase the carbon consumption by 27% in response to a doubling in present day CO_2 (Riebesell et al., 2007). The evidence from experiments on natural plankton communities is equivocal, with examples of both increasing and decreasing C:N ratios (Hutchins et al., 2009). In a model study, the hypothesized effect of enhanced organic carbon export due to elevated C:N ratio resulted in a moderate increase in oceanic CO_2 uptake (a cumulative value of 35 Pg C by 2100) and a fifty percent increase in the extent of subsurface low-oxygen zones in the tropical ocean (Oschlies et al., 2008). In addition, increased production of extracellular organic matter under high CO_2 levels (Engel, 2002) may enhance the formation of particle aggregates (Engel et al., 2004; Schartau et al., 2008) and thereby increase the vertical flux of organic matter (Riebesell et al., 2007; Arrigo, 2007), which may also affect nutrient availability for phytoplankton in surface waters.

Ocean acidification has the potential to alter the marine nitrogen cycle which controls much of primary production in the sea. Laboratory experiments with the nitrogen-fixing cyanobacterium *Trichodesmium* revealed an increase in both carbon and nitrogen fixation with increasing pCO_2 (Barcelos e Ramos et al., 2007; Hutchins et al., 2007; Levitan et al., 2007; Kranz et al., 2009). Since *Trichodesmium* is a dominant species in large parts of the nutrient-poor tropical and subtropical oceans, this response has the potential to increase the reservoir of bioavailable nitrogen in the surface layer of these areas. These areas of the ocean are predominantly nitrogen-limited; therefore, an increase in nitrogen fixation would provide additional new nitrogen in low-nutrient subtropical regions and would lead to increased primary production and carbon fixation. The actual increase in nitrogen fixation, however, could be limited by phosphorus or iron supplies. A strong positive relationship between nitrogen fixation and rising CO_2 has also been observed for cultured *Crocosphaera*, a nitrogen-fixing unicellular cyanobacterium, under iron-replete conditions but not under iron limited conditions (Fu et al., 2008), but another nitrogen-fixing cyanobacterium, *Nodularia spumigena*, showed the opposite response (i.e., reduced growth rate and nitrogen fixation rate at elevated CO_2; Czerny et al., 2009).

These effects on calcification, photosynthesis, nitrogen fixation, and other processes will likely lead to shifts in the planktonic community as some species fare better than others under acidification. However, no consistent responses have been obtained in experiments concerning the effect of ocean acidification on plankton community composition. In one experiment with a phytoplankton community dominated by microflagellates, cryptomonads, and diatoms, only the diatom *Skeletonema costatum* responded to elevated CO_2 by increased growth rate (Kim et al., 2006). A similar shift in phytoplankton species composition from *Phaeocystis* to diatom dominance occurred in another shipboard incubation experiment (Tortell et al., 2002). In contrast, a remarkable resilience of the enclosed plankton communities to seawater acidification was observed in a series of mesocosm CO_2 enrichment experiments: no significant differences between CO_2 treatments were observed for phytoplankton composition and cell cycle, inorganic nutrient utilization and nutrient turnover, bacterial abundance and diversity, micro-zooplankton grazing and copepod feeding and egg production (Riebesell et al., 2008). While shifts in planktonic community composition could theoretically affect higher trophic levels, no experimental results exist to confirm these predictions.

Another important consideration is the possible interactive effects of climate change and acidification such as the warming of surface waters and reduced nutrient availability. Similarly, ocean microbes produce and destroy a number of trace gases that are important for atmospheric chemistry and climate besides CO_2 and O_2. For example, nitrous oxide (N_2O), a powerful greenhouse gas, is a by-product of both nitrification and denitrification and its marine production might thus be affected by acidification. Another important trace gas produced in the oceans is dimethylsulfide (DMS), which serves as a precursor for atmospheric sulfate aerosols that nucleate cloud droplets and cool surface temperatures. Mesocosm experiments at elevated CO_2 (Vogt et al., 2008; Wingenter et al., 2007; Hopkins et al., 2010) have shown both positive and negative responses in dissolved DMS responses with both small decreases. In this way, changes in the microbial community composition and activity triggered by ocean acidification may act as a feedback on climate change.

4.3 COASTAL ECOSYSTEMS

Coastal ocean ecosystems include a variety of benthic habitat types, including seagrass beds, kelp forests, tidal wetlands, mangroves, and others. They represent some of the most productive marine ecosystems that support numerous finfish and shellfish fisheries, both managed and cultured. Humans rely on coastal ecosystems for commerce, recreation, protection from storm surges, and a suite of other services; however, there is also a great deal of anthropogenic impact on coastal habitats. This section does not attempt to review all of the possible impacts of acidification on the various types of coastal ecosystems. Rather it highlights some general concerns, particularly for important coastal species and functions such as commercially-important fishery species and ecosystem engineers. Ocean acidification may affect coastal ecosystems in a variety of ways. It can directly impact the growth and survival of coastal organisms, particularly in sensitive reproductive and early developmental stages. It can also affect growth and survival indirectly by altering food web dynamics and nutrient cycling. It is also likely to affect important coastal ecosystem engineers that create habitat.

A major focus of recent studies has been on the potential effects of ocean acidification on the early life history of various species. For many coastal benthic calcifiers, including commercially-important species, reproduction and early development appear to be particularly sensitive to acidification (Kurihara, 2008). Reduced growth and calcification rates, and in some cases even shell dissolution and mortality, have been reported for larval and juvenile stages in a number of bivalve species: the bay scallop *Argopecten irradians* (Talmage and Gobler, 2009), the hard clam *Mercenaria mercenaria* (exposed to sediments that were undersaturated with respect to aragonite; Green et al., 2004, 2009), the soft-shell clam *Mya arenaria* (Salisbury et al., 2008), the Mediterranean mussel *Mytilus galloprovincialis* (Kurihara et al., 2008a), the Sydney rock oyster *Saccostrea glomerata* (Watson et al., 2009), the Pacific oyster *Crassostrea gigas* (Kurihara et al., 2007), and the Eastern oyster *Crassostrea virginica* (Miller et al., 2009). Interestingly, Miller et al. (2009) did not see similar effects on the Suminoe oyster, *Crassostrea ariakensis*, indicating a species-specific response that could lead to shifts in community composition. Hence, these comparative studies did find that some species were more tolerant of high CO_2 conditions. Negative effects of acidification have also been seen in the early development of non-bivalve species such as the European lobster *Homarus gammarus* (Arnold et al., 2009), the Pacific shrimp *Palaemon pacificus* (Kurihara, 2008), and the sea urchin *Echinometra mathaei* (Kurihara and Shirayama, 2004). In contrast, juveniles of American lobster (*H. gammarus*) and the blue crab (*Callinectes sapidus*) showed elevated rates of calcification at very high pCO_2 levels (Ries et al., 2009).

Many studies have also shown negative effects on adult growth and survivorship of these and other coastal benthic species (e.g., Gazeau et al., 2007; Kurihara et al., 2008b; Ries et al. 2009). There were mixed responses—increasing, decreasing, parabolic, and no change in calcification rates—to decreasing saturation state in the eighteen benthic coastal species studied by Ries et al. (2009). It is not known whether positive or negative changes in calcification in these organisms would affect their lifelong productivity, growth, and fitness. Impacts on many other species not yet studied are likely.

Indirectly, acidification may affect the productivity and composition of some coastal ecosystems by affecting the key species at the base of coastal food webs. As noted previously, several calcifying planktonic species are sensitive to seawater pH and carbonate chemistry changes and can be important prey species in coastal ecosystems (e.g., pteropods may be important prey in salmon diets, Armstrong et al., 2008). In addition, the planktonic larvae of many species are also prey items and, as previously discussed, may be negatively affected by acidification. Therefore, coastal organisms that are not directly susceptible to the effects of acidification may indirectly be affected through trophic interactions.

Many coastal habitats depend on ecosystem engineers to build and maintain structures that provide critical habitat for other organisms, including oyster reefs, kelp forests, and seagrass beds. Oysters have already been discussed as species that will likely be negatively affected by acidification. On the other hand, research has shown increased growth of seagrass (Figure 4.1) with increased CO_2 (Zimmerman et al., 1997). It is probable that an increase in total seagrass area will lead to more favorable habitat and conditions for associated invertebrate and fish species (Guinotte and Fabry, 2009).

Coastal ecosystems exhibit naturally high variability in pH and seawater chemistry due to biological activity, freshwater input, upwelling, atmospheric deposition, and other factors. They are also subject to a diversity of stresses caused by human activities, such as organic matter and nutrient inputs, pollution by toxic organic compounds and metals, acid rain, sea level rise and other climate change effects, and overfishing. The effects of ocean acidification on coastal ecosystems may be small relative to the effects of these natural and human-induced stresses. But in some instances, acidification may act synergistically with other factors (Figure 4.2). For example, coastal upwelling is a natural phenomenon that brings deep water to the surface; this water is often undersaturated with respect to calcium carbonate. However, further acidification of these upwelled waters by anthropogenic CO_2 uptake may be increasing the intensity and areal extent of these "corrosive" events (Feely et al., 2008). Increased temperature due to climate change is another stressor that is likely to interact with acidification; for example, temperature has been shown to act synergistically with acidification in the development of the Sydney rock oyster (Parker et al., 2009). Another likely interaction is that of increased nutrients and acidification. For example, in kelp forests, it is predicted that local nutrient pollution and increased CO_2 will enhance the growth of filamentous algae species while simultaneously decreasing calcifying macroalgae that serve as the understory of kelp forests, thus allowing for a shift from kelp forests to filamentous turf mats (Russell et al., 2009).

Another example of a potential synergism is the interaction between acidification and low oxygen (i.e., hypoxic) or no oxygen (i.e., anoxic) "dead zones." The decomposition of organic matter near the bottom in shallow coastal waters increases the ambient CO_2 concentration and decreases the oxygen concentration and pH. This natural phenomenon can be exacerbated by anthropogenic inputs of organic waste and algal nutrients, resulting in dead zones. But in regions that are only hypoxic, the low oxygen and the high CO_2 tend to act in concert to make respiration difficult for a number of aerobic organisms. It is possible that a further increase in CO_2 caused directly or indirectly by acidification could increase the intensity or spatial extent of the hypoxic and anoxic events. Examples of ecosystems where this could occur is along many highly productive coastal upwelling zones around the world, such as the eastern Pacific, the Arabian Sea, and along northern and southern west Africa.

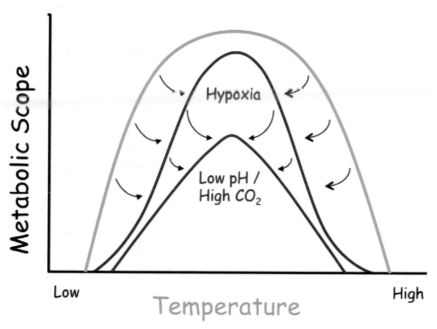

FIGURE 4.2 Specific combinations of environmental factors affect animal performance in ways that can narrow the range of performance for any given factor. These windows of performance (modified from Pörtner and Farrell, 2008) for organisms can be measured along environmental gradients such as temperature. In the example illustrated in the graph, an organism may have a relatively broad temperature tolerance (green line, from low to high), but this tolerance may only be observed under oxygenated conditions and normal seawater pH. Both low oxygen (hypoxia) and lower pH/high CO_2 conditions could not only reduce the overall organismal performance, but also could narrow the temperature range under which this organism could survive. Hence, for some organisms, ocean acidification would restrict the habitable range of temperature and reduce the performance range (the metabolic scope which represents the maximium minus the minimum metabolic rate).

There, as previously discussed, the natural cycle in acid-base chemistry resulting from seasonal upwelling is amplified by penetration of anthropogenic CO_2 into the upwelled water. The ambient flora and fauna, particularly benthic organisms, may well be affected by annual exposure to acidic and, in some cases, corrosive hypoxic water.

Depending on the differential tolerances of organisms to changes imposed by acidification, there are likely to be shifts in community composition or productivity of the various ecosystems. However, existing research in coastal ecosystems, as is the case with other ecosystems, has been focused on individual organisms, not on the population, community, or ecosystem levels. Consequently, it is unknown whether populations sensitive to changes in ocean chemistry will be able to adapt through behavioral or physiological changes. For example, populations with individuals possessing genetic variations that tolerate the expected changes in ocean chemistry may result in higher survival or reproductive success because of more-rapid-than-expected adaptation to the new conditions.

It is not known whether coastal ecosystems that do not currently experience natural hypoxic and low pH events are less susceptible to incremental shifts in regional ocean chemistry due to ocean acidification. Areas along the U.S. eastern seaboard, the Gulf of Maine, and others have weaker oxygen minimum zones and higher pH waters along coastal zones. Organisms inhabiting these ecosystems may tolerate larger shifts in ocean chemistry caused by ocean acidification than those in ecosystems overlying more hypoxic upwelling waters, but this hypothesis requires study. Hypoxic dead zones caused by anthropogenic sources have been observed in most urbanized coastlines of the world, regardless of regional oceanography. These events, also accompanied by low pH, may indicate that most coastal ecosystems are sensitive to extreme eutrophication events.

4.4 DEEP SEA, INCLUDING COLD-WATER CORALS

Acidification of the deep ocean will occur more slowly than in surface seawater. But its ecological effects may nonetheless be severe because of the assumed greater sensitivity of the deep biota. Deep-sea organisms live in a cold, dark environment with low nutrient inputs and reduced reliance on visual interactions between predator and prey. These organisms generally grow slowly and have lower metabolic rates than comparable taxa living in warmer surface waters (Seibel and Walsh, 2001, 2003; Goffredi and Childress, 2001; Seibel et al., 1997; Gage and Tyler, 1991; Pörtner et al., 2004). In animals, slow metabolism typically corresponds to a low capacity for gas exchange (i.e., oxygen transport and CO_2 release) and reduced enzyme function, including those linked to acid-base regulation (Seibel and Drazen, 2007; Melzner et al., 2009). For example, a logarithmic decrease in passive pH buffering ability with depth has been measured in highly active pelagic predatory cephalopods (Seibel and Walsh, 2003), indicating increasing vulnerability to acid-base disturbance with depth. The environmental stability of the deep sea over long time scales is also postulated to have reduced the tolerance of deep-sea species to environmental extremes through the loss of more tolerant genotypes (Dahlhoff, 2004), thereby decreasing the potential for adaptation to future ocean acidification.

Experimental studies with deep-sea organisms are obviously difficult and very few provide direct information on their sensitivity to acidification. In experiments performed on the abyssal floor off central California, low rates of survival of deep benthic organisms were observed after exposure to a modest decrease in pH (-0.2 units) near pools of liquid CO_2 (Barry et al., 2003, 2005; Thistle et al., 2005; Fleeger et al., 2006). In contrast, deep-sea fish and cephalopods survived month-long exposure to mildly acidic waters during these experiments (Barry and Drazen, 2007), although related physiological studies indicate that respiratory stress (impaired oxygen transport) is likely for deep-living cephalopods exposed to low pH waters (Seibel and Walsh, 2003). In other experiments, deep sea crabs were much less able to recover from short-term exposure to very high CO_2 than shallow dwelling crabs and this effect was amplified at low oxygen concentrations (Pane and Barry, 2007).

Some likely consequences of future ocean acidification in deep-sea waters can be inferred from organisms inhabiting hydrothermal vent and cold seep environments, which often (but not always) have low pH levels. Echinoderms and some other calcifying taxa are generally absent from hydrothermal vents (Grassle, 1986) and cold seeps (Sibuet and Olu, 1998), presumably as a result of the low ambient pH or other stressful environmental factors. For example, high concentrations of toxic metals (e.g., cadmium, silver, strontium, barium, and others) in vent effluent at some sites (Van Dover, 2000) may limit distribution of some fauna. Other vent and seep taxa thrive, in spite of high CO_2 levels, and in some cases exploit the energy-rich conditions in these environments to sustain anomalously high rates of growth (Barry et al., 2007; Urcuyo et al., 2007). Adaptations promoting success for some animals at vent and seep habitats are likely to have evolved over long periods; it remains unknown whether more typical deep-sea animals are capable of adapting to future changes in deep ocean chemistry caused by acidification.

A unique habitat type in the deep sea that deserves particular attention is cold-water coral communities. Cold-water corals, also known as deep-water or deep-sea corals, form ecosystems that are in some ways the deep-water counterparts of tropical coral reefs. Cold-water coral reefs (or bioherms) are also founded on the accumulation of calcium carbonate, providing the structural framework for these biodiverse ecosystems that serve as habitat for a range of organisms, including commercially important fish species (Freiwald et al., 2004; Roberts et al., 2006). The primary reef-building species are stony corals that lack zooxanthellae, the symbiotic algae common in shallow, tropical species. Cold-water coral ecosystems occur globally in darker, colder waters than their tropical counterparts, from depths as shallow as 40 m to greater than 1,000 m (Freiwald, 2002; Freiwald et al., 2004).

As with tropical coral reefs, the main concern for cold-water corals with respect to ocean acidification is the effect on calcification rates for key reef-builders. The geographic distribution of cold-water coral communities suggests that they are limited to waters supersaturated with respect to their predominant skeletal mineralogy aragonite (Guinotte et al., 2006). With expected shoaling of the aragonite saturation horizon, many of these communities may become exposed to waters corrosive to coral skeletons. However, it is unclear whether it is the species or the structures they construct (or both) that are limited by the saturation horizon. Calcification rates in the cold-water species *Lophelia pertusa* were reduced by an average of 30 and 56% when pH was lowered by 0.15 and 0.3 units relative to ambient conditions, respectively (Maier et al., 2009), but despite this response, calcification rates in this species did not stop completely even in aragonite-undersaturated conditions. It must be noted that this is the only study on the response of a cold-water coral species to ocean acidification.

Deep-sea coral communities are also abundant and ecologically significant on thousands of seamounts throughout the world ocean that could be affected by ocean acidification. Seamounts—undersea mountains that rise from the abyssal plain but do not breach the surface—number about 100,000 worldwide (Figure 4.3). The coral- and sponge-dominated assemblages found near the peaks of seamounts depend nutritionally on suspended organic debris sinking from sunlit surface waters and form important habitat for deep-sea fisheries, including orange roughy, alfonsino, roundnose grenadier and Patagonian toothfish (Clark et al., 2006). Corals that dominate seamount assemblages include stony corals (scleractinians), black corals (Antipatharians), and octocorallians, including sea fans (gorgonians). Waters around seamounts and throughout the deep-sea are naturally more acidic than those found in shallower depths because of the accumulation of carbon dioxide from the respiration of deep-sea organisms. This effect is greatest in areas such as the Northeast Pacific Ocean. Mixing of anthropogenic carbon dioxide into the deep-sea will make these waters even more acidic. Aragonitic corals are much less abundant in the more acidic waters of the Pacific Basin (Roberts et al., 2006), and most species appear to be limited in distribution by the depth of the existing saturation horizon for aragonite, as shown by the strong reduction in the abundance and diversity of scleractinians below this boundary (Guinotte et al., 2006; Cairns, 2007). For seamounts with summits that are more than a few hundred meters below the surface, especially in the Pacific basin where waters are corrosive or nearly so to aragonite, the most common corals are calcitic, including the gorgonians, which often dominate as habitat-forming species. For example, the bubble-gum coral (*Paragorgia* sp.; Figure 4.1) is a common coral found worldwide

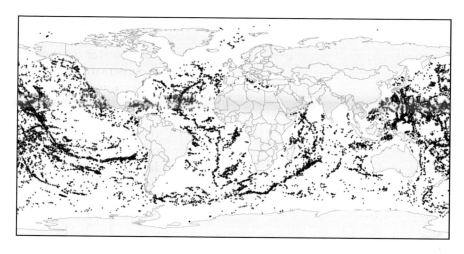

FIGURE 4.3 Global dataset of more than 30,000 potential seamounts (Kitchingman and Lai, 2004; www.seaaroundus.org). Estimates of the total number of seamounts in the world ocean varies greatly depending upon the resolution of bathymetric data available and analytic methods used. The abundances of deep-sea corals on seamounts are correlated closely with the aragonite and calcite saturation horizons (Guinotte et al., 2006).

on seamounts, and can reach at least 3 m in height (Mortensen and Buhl-Mortensen, 2005). Like aragonitic corals, gorgonians and other calcitic corals are likely to be affected by changes in calcite saturation with depth, though protective coverings and tissues may provide some protection from carbonate dissolution.

Seamount coral communities are highly vulnerable to anthropogenic disturbances. Growth rates of deep-sea corals are known to be low, with longevity estimates ranging from at least decades to centuries (e.g., Andrews et al., 2002; Clark et al., 2006), with at least some species living more than 1,000 years. Longevity estimates of some corals from ~500 m depth off the Hawaiian Islands were estimated at 2,742 y (*Gerardia* sp.) and 4,265 years (*Leiopathes* sp.) (Roark et al., 2009). The slow growth and long recovery time of seamount coral communities put them at greater risk for damage from human activities, including ocean acidification. Considering the expected rapid shoaling of the calcite and aragonite saturation horizons with future ocean acidification and the observed relationship between coral distributions and existing saturation horizons, deep-sea coral communities on seamounts or bioherms are likely to be impacted.

4.5 HIGH LATITUDES

High latitude waters of the Arctic and Southern oceans are very productive and support diverse pelagic and benthic communities. Some
of the richest and most heavily exploited fishing areas in the world are
located in high latitude waters, including the northern Bering, Chukchi,
and Barents Seas in the Arctic and a krill fishery in the Southern Ocean
(Dayton et al., 1994). About half of the U.S. domestic fish catch by biomass tonnage is landed in Alaska (Fisheries Economics of the U.S., 2008[3]).
Many protected and endangered marine mammals and seabirds also roam
high latitude waters. High biodiversity cold-water coral habitats can be
found in the high latitudes, including the "coral gardens" off the Aleutian
Islands (discussed in further detail in section 4.4). Yet high latitude organisms are not as well studied as those in lower latitudes and the effects of
ocean acidification on polar and subpolar marine life and ecosystems are
largely unknown.

Like many other ecosystems, the most likely threat that acidification
poses in the high latitudes is to planktonic calcifiers. In the subarctic
Pacific, pteropods can be important prey of juvenile pink salmon, accounting in some years for >60% by weight of their diet (Armstrong et al.,
2005). When exposed to the level of aragonite undersaturation expected
to occur by the year 2100 (see Figure 2.10), thecosomatous pteropods
showed visual evidence of reduced calcification (Comeau et al., 2009; Orr
et al., 2005). If thecosomatous pteropods cannot adapt to living continuously in seawater that is undersaturated with respect to aragonite, their
ranges will contract to shallower depths and lower latitudes that have
higher carbonate ion concentrations. The possible exclusion of pteropods
from high latitude regions would impact the downward organic carbon
flux associated with pteropod fecal pellets (Thibault et al., 1999; Collier
et al., 2000) and remove a major source of calcium carbonate in such
regions (e.g., Bathmann et al., 1991; Honjo et al., 2000; Gardner et al.,
2000; Accornero et al., 2003; Tsurumi et al., 2005). Similarly, if foraminifera
densities decrease in some high latitude areas where they are currently
abundant (e.g., subarctic Pacific), calcium carbonate export to the ocean
interior will be reduced, which would in turn decrease the potential of
foraminiferal tests to act as ballast in the transport of organic carbon to
the deep sea (Schiebel, 2002; Moy et al., 2009). As in other regions, ocean
acidification could also alter the species composition of primary producers and rates of photosynthesis through pH-dependent speciation of
nutrients and metals (Zeebe and Wolf-Gladrow, 2001; Byrne et al., 1988;
Shi et al., 2009; Millero et al., 2009).

---

[3] http://www.st.nmfs.noaa.gov/st5/publication/fisheries_economics_2008.html

Polar benthic communities may also be affected by acidification. Although there are major differences in the modern biota and structure of benthic communities in the Arctic and Southern Ocean that reflect the distinct topography and evolutionary history of the polar habitats, there may be similar vulnerabilities in the two systems. Polar invertebrates tend to have low metabolic rates and slow growth rates. In addition, high latitude benthic (and some planktonic) invertebrates can have long generation times compared to warmer water taxa, providing them fewer opportunities to evolve effective adaptations to cope with seawater that will be progressively depleted in carbonate ion concentration and corrosive to calcium carbonate minerals in the coming decades (Orr et al., 2005; Bates et al., 2009; Olafsson et al., 2009). Calcifying macroalgae and marine invertebrates, including cold-water corals, sea urchins, and molluscs, make up significant components of the rich benthic communities in high latitudes, and these are thought to be at risk with increasing ocean acidification.

The aragonite saturation state of seawater provides a clear geochemical threshold when seawater becomes undersaturated with respect to aragonite. While many studies indicate that calcification correlates with the calcium carbonate saturation state of seawater, biological thresholds of the calcification response to ocean acidity may be species-specific. Such differential responses of species to rising ocean acidity may result in competitive advantages that could drive the reorganization of planktonic and benthic ecosystems, thereby affecting food webs, fisheries, and many ecological processes. The high latitudes will be the first ocean regions to become persistently undersaturated with respect to aragonite as a result of anthropogenic-induced acidification (Figure 2.10). Thus, these ecosystems are natural laboratories in which to test many hypotheses on the impacts of ocean acidification and other stressors, particularly those induced by global warming.

Many polar and subpolar ecosystems are undergoing rapid change owing to global warming. The reduction in sea ice, freshening of seawater, and increasing ocean and air temperatures are forcing major ecological shifts in polar regions of both hemispheres. The western shelf of the Antarctic Peninsula is the fastest warming region on earth, with rates of temperature increase nearly five times the global average rate over the past century (Ducklow et al., 2007). Warming sea temperatures may allow shell-crushing crabs to invade the shelf benthos surrounding Antarctica, with significant consequences for benthic organisms that have evolved in the absence of such predators (Aronson et al., 2007). Since the Eocene, cold temperatures have prevented crabs from invading Antarctic shelves; however, king crabs are moving up the western Antarctic continental slope (Thatje et al., 2005) and should they arrive on the continental shelves, the weakly calcified shells of Antarctic echinoderms and molluscs—further stressed by acidification—would provide little defense from these predators. A change from arctic to subarctic conditions is underway in the northern Bering Sea, and poleward displacement of marine mammals has coincided with a reduction in benthic prey, an increase in pelagic fish, and reduced sea ice (Grebmeier et al., 2006). Again, acidification impacts on prey species could further exacerbate food web changes caused by changing climate conditions. In both hemispheres, the observed regional changes are expected to affect broader areas of the Arctic and Southern Oceans, respectively, in future decades. In addition to warming temperatures, retreat of sea ice and increasing species invasions, high latitude regions, particularly in the north, are subject to heavy fishing pressure which is an additional stressor for these ecosystems.

4.6 LESSONS FROM THE GEOLOGIC PAST

Evidence from the geologic record indicates that the Earth previously experienced periods of high atmospheric CO_2 which also changed ocean chemistry. Studies of past ocean chemistry and coincident changes in marine ecosystems may provide insight into the potential impacts of ocean acidification today and in the future.

Approximately 55 million years ago, a large release of carbon into the oceans changed the Earth's climate and ocean chemistry, an event called the Paleocene-Eocene thermal maximum (PETM). Atmospheric CO_2 and global temperature spiked upward and then slowly recovered over a period of more than 100,000 years (Kennett and Stott, 1991; Pagani et al., 2006; Zachos et al., 2001). The evidence from the isotopic compositions of carbon ($\delta^{13}C$) and oxygen ($\delta^{18}O$) in $CaCO_3$ in deep ocean sediments indicate that the release of carbon was relatively rapid (~10,000 years) though the exact duration of the release event is not well constrained by the sedimentary data. The $\delta^{13}C$ of surface-dwelling plankton appeared to change instantaneously, while benthic foraminifera recorded transitional $\delta^{13}C$ values, as if the atmospheric CO_2 changed on a time scale shorter than the circulation time of the ocean (Thomas et al., 2002), which today takes about 1,000 years. However, a longer CO_2 release time of 10,000 years is suggested by the sedimentary time scale based on orbital variations (Lourens et al., 2005). The oxygen isotopic composition of the $CaCO_3$ indicates that intermediate-depth ocean, and presumably the Earth's surface, warmed in concert with the carbon release. Both temperature and CO_2 gradually returned to their initial, steady values (Lourens et al., 2005). The recovery to initial conditions of carbon and oxygen occurred on a time scale, over 100,000 years, comparable to the silicate weathering thermostat mechanism for regulating atmospheric CO_2 (Berner and

Kothavala, 2001), a further indication that CO_2 played a role in the spike in global temperature.

Deep sea sediments from the PETM show extensive dissolution of $CaCO_3$ (Zachos et al., 2005), consistent with an elevation in atmospheric CO_2. Somewhat puzzlingly, the extent of $CaCO_3$ dissolution differs greatly between the Atlantic and Pacific basins during that time (Zeebe and Zachos, 2007), possibly the result of regional anoxia events that would reduce mixing of surface sediments. Nonetheless, a number of factors limit the utility of the PETM as an analog for the detailed effects of acidification on the biota and carbon cycle of the ocean. First, the amount of carbon released is not well constrained because the exact source is unknown, and the magnitude of carbon isotope excursions in different carbon isotopic records vary by roughly a factor or two, with larger excursions typically found in soil carbon records than in deep sea sediments. Second, the magnitude of the ocean pH excursion is also unclear because it is dependent on whether the CO_2 release was faster or slower than the $CaCO_3$ neutralization time scale.

The PETM was marked by the extinction of $CaCO_3$-producing foraminifera that live on the sea floor, perhaps in response to acidification or alternatively as a result of anoxia in the deep sea. There was not a comparable extinction in shallow-water species such as mollusks, but the occurrence of weakly calcified planktonic foraminifera may indicate changes in carbonate ion concentration in surface waters. A decrease in productivity or diversity, which would be relevant to humankind in the future, is difficult to gauge from the fossil record.

The impact of a comet or asteroid at the boundary between the Cretaceous and the Tertiary periods (also known as the K/T boundary), which occurred 65 million years ago and is responsible for the extinction of the dinosaurs, may have also perturbed the pH of the ocean. In this event, the impact fireball caused the oxidation of atmospheric nitrogen to nitric acid (D'Hondt and Keller, 1991) and produced sulfuric acid from the calcium sulfate enriched carbonate structures at the point of impact (D'Hondt et al., 1994). The atmospheric deposition of nitric and sulfuric acids likely only affected the pH of surface waters which would have recovered ambient pH relatively quickly as they mixed with deeper water. The impact also released large quantities of dust and aerosols that would have darkened the skies and cooled Earth's atmosphere. As in the PETM, calcifying organisms suffered greater extinction rates than organisms that do not produce $CaCO_3$, but the ecological responses that can be reconstructed could have been the result of the collapse of photosynthesis from the darkened skies, or disruption of other geochemical factors, in addition to or instead of changes in ocean pH.

The largest extinction event in Earth's history took place 251 million years ago at the boundary between the Permian and Triassic periods (Knoll et al., 1996). The cause of this event is speculative; possibilities include the impact of a large object (such as a meteor), extensive volcanism, ocean anoxia, or release of methane from methane hydrates. Analysis of the correlations between extinction patterns and physiology suggest that elevated CO_2 levels might have played a role, but the duration over which this extinction occurred is unknown.

These three geological events give general support to current concerns about ocean acidification, particularly related to the possibility that calcifying organisms may decrease or even disappear as a result of increasing CO_2. However, the severity of the perturbations and their durations are not known with enough accuracy to determine their similarity to conditions resulting from anthropogenic CO_2 emissions. As a consequence, responses of marine ecosystems to the ongoing increase in CO_2 may not be analogous to the changes in biological diversity associated with events in the deep past. Further development of proxy measurements, such as the use of boron isotopes to estimate ocean pH changes, could provide additional information on the rate and extent of changes in ocean CO_2 and pH during these past climatic events.

4.7 BIODIVERSITY, THRESHOLDS, AND MANAGING FOR CHANGE

Regardless of the ecosystem, there is a concern that ocean acidification, along with other stressors, will reduce the biodiversity (i.e., species richness) of marine ecosystems through species extinctions, with potentially important consequences. Changes in species' abundances, either directly due to the tolerance or intolerance of species to ocean acidification, or indirectly through changes in competitive interactions and trophic linkages, are very likely in the future.

Depending on the sensitivities of species, ocean acidification may result in extinctions that reduce the biodiversity of marine communities. Very little information is available on the effects of ocean acidification on biodiversity, but studies in areas where the water is naturally high in CO_2 may provide some indication of the types of changes that could occur with global ocean acidification. For example, studies of species composition in the vicinity of CO_2-rich volcanic vents in the Mediterranean Sea suggest that acidification will reduce the biodiversity of shallow, marine benthic communities (Hall-Spencer et al., 2008). High biodiversity in marine ecosystems is generally considered to enhance the stability of ecosystems through "functional redundancy" or "species complementarity." In other words, when biodiversity is high, there are many species serving similar ecological roles. Reduced ecosystem biodiversity due to the loss of species increases the dependence of the ecosystem on the services (e.g., prey or predatory rates) provided by the remaining similar species. If key trophic linkages are lost (e.g., an intermediate consumer guild is reduced severely), food web integrity may be compromised, energy flow may be impaired, and significant changes in ecosystem structure and function become likely—an ecological tipping point or threshold has been broached that can lead to a catastrophic change in an ecosystem. These "regime shifts" can move an ecosystem from one stable state to an entirely different state.

Many ecosystems have been demonstrated to undergo regime shifts to alternative ecological states (Scheffer et al., 2001). Analyses of previous regime shifts in both terrestrial and marine ecosystems (e.g., rangelands (Briske et al., 2005), lakes (Carpenter et al., 1999), coral reefs (Norström et al., 2009), open ocean (Overland et al., 2008)) show that they were rarely predicted, and many appeared to be triggered by relatively small events (van Nes and Scheffer, 2004). The growing body of literature now illustrates that the underlying cause for regime shifts is a decrease in ecosystem resilience (Folke et al., 2004; Scheffer et al., 2001). Resilience can be defined as "the amount of change or disturbance that a system can absorb before it undergoes a fundamental shift to a different set of processes and structures" (West et al., 2009). In many regime shifts, once an ecological threshold has been passed, the driver of the change must be reversed to levels far beyond where the shift occurred before the system shifts back to its original state. Regime shifts are likely within those marine ecosystems that experience stress from ocean acidification, either directly (e.g., through elimination of one or more species) or indirectly (e.g., alteration of the physical environment, such as dissolution of substrate), and particularly in combination with other stressors. Ecosystems degraded by acidification also may become more sensitive to other human and climate change stressors beyond ocean acidification.

As stated by Overland et al. (2008) "our current understanding of regime shifts is not a deterministic one, and while one can discuss amplitudes and mean duration of regimes, we cannot predict their precise timing other than to say that they will be a main feature of future climate and ecosystem states." Nonetheless, developing methods for detecting, and in some cases even predicting or managing, an ecosystem's approach toward a tipping point or critical threshold has received increasing attention (e.g., de Young et al., 2008; Scheffer et al., 2009). Multiple techniques for identifying regime shifts are now available, but only after they have occurred (Andersen et al., 2009; Carpenter et al., 2008). Recent evidence, suggests that complex systems (including ecosystems) may exhibit certain "symptoms" prior to a regime shift (Scheffer et al., 2009), such as:

(1) a "critical slowing down" of the dynamics which would be expressed as a slower recovery from small perturbations, increased auto-correlation (Dakos et al., 2008), or a shift of variance power spectra toward lower frequencies (Kleinen et al., 2003; Dakos et al., 2008),

(2) notably increased variance (Carpenter and Brock, 2006),

(3) greater asymmetry in fluctuations (Guttal and Jayaprakash, 2008), and

(4) in benthic communities, a breakdown of scaling rules for spatial patterns (Rietkerk et al., 2004).

Recent progress has been made toward attributing ecological shifts, particularly in terrestrial systems, to climate change (Rosenzweig et al., 2008). A major challenge in ocean acidification research is how to attribute ecological shifts to forcing from ocean acidification. In the field, ocean acidification rarely, if ever, will be the only driver of change. Climate change is simultaneously causing changes in temperature, circulation patterns, and other phenomena, so that attribution of changes (or at least part of the change) to ocean acidification will be difficult. In coral reefs, for example, whether the loss of corals is due to rising temperature or from ocean acidification may have little relevance in the overall impact on the ecosystem (loss of corals impacts the base function of the ecosystem). But systems where species are differentially impacted by temperature and/or ocean acidification may exhibit clear signs as to which factor is likely to cause a major ecological shift. Analyses of changes in food webs supporting fisheries, for example, reveal patterns that indicate whether the drivers of that change lie near the base of the food chain or at the top (Frank et al., 2007).

Management of ecological systems for climate change has focused primarily on adaptations that maintain or increase ecosystem resilience (West et al., 2009). The most common recommendation for maintaining resilience is to limit local to regional stressors such as land-based pollution, coastal development, overharvesting, and invasive species. Ecosystems with high biodiversity and/or redundancy of functional groups (e.g., several species fill the role of algal grazers) tend to be more resilient, and recover more quickly following a perturbation, which suggests that managing for biodiversity is a logical means of sustaining ecosystems (Palumbi et al., 2009). Resilience of some stocks to overfishing, for example, appears to be related to warmer regions with greater species richness (Frank et al., 2006; Frank et al., 2007). This suggests that different strategies may be necessary for maintaining resilience across different ecosystems.

5

Socioeconomic Concerns

Marine ecosystems provide humans with a broad range of goods and services, including seafood and natural products, nutrient cycling, protection from coastal flooding and erosion, recreational opportunities, and so-called "nonuse values" such as the value that people ascribe to continued existence of various marine species. As outlined in previous chapters, many of these goods and services may be affected by ocean acidification (Cooley et al., 2009), and measuring and valuing these impacts on society can help guide policy and management decisions. For example, understanding the overall economic impact of ocean acidification can enhance the discussion of national and international climate change mitigation options (e.g., reducing CO_2 emissions). However, it may be even more useful to provide information that empowers stakeholders and enables decision makers to respond constructively to ocean acidification. To provide such information, one must determine who will be affected, when, and by how much, and how those impacts might be anticipated, prepared for, or mitigated.

As with the ecological effects, the economic implications of ocean acidification are presently not well understood. Potential economic harms as well as opportunities are only now being identified (Cooley and Doney, 2009). This chapter first presents a brief,[1] general discussion of how the impacts can be measured and valued. It then considers three sectors—

---

[1] Holland et al., 2010 provide a more detailed discussion of how economic evaluation frameworks and economic modeling and valuation methods can be applied to evaluating impacts on ecosystems and ecosystem services.

fisheries, aquaculture, and tropical coral reef systems—for which socioeconomic impacts appear most probable based on currently available data
and which have attracted the most public attention and concern (e.g., Pew
Center for Global Climate Change, 2009).

5.1 EVALUATING IMPACTS ON SOCIETY

Economic methods and models can be used to estimate how net benefits to society may be affected by expected changes in marine ecosystems
due to ocean acidification (see previous chapters) and to assess the value
of responses to those changes. Economic analysis can provide information on how best to reduce economic harm or to capitalize on opportunities brought on by ocean acidification. While economic values are not
the only, or even necessarily the most important, criteria for informing
decisions on responses to ocean acidification, they do provide a means
to compare alternative uses of society's resources with a framework that
relates value to human welfare in terms of individuals' assessments of
their personal well-being (Bockstael et al., 2000). The strong theoretical
and empirical foundation of economics enables the measurement of quantitative, logically consistent, and directly comparable measures of human
benefits and costs, whether realized through organized market activity
or outside of markets. Like other natural or social sciences, the accuracy
of these and other economic predictions is generally highest for small
(marginal) or localized changes. As one moves further from the current
condition, expected accuracy declines. Hence, it may not be practical or
meaningful to quantify the value of the loss or restructuring of an entire
ecosystem, but it is possible to quantify the value of discrete changes in
the ecosystem services relative to a specific baseline.

Economic valuation methods can be applied both to market goods
(e.g., seafood) and non-market goods (e.g., protection from coastal flooding and erosion). Many of the economic effects of ocean acidification will
be on ecosystem services that are not traded in markets but still have
substantial economic value. A variety of different non-market valuation
methods can be used to quantify these benefits, each suited to the measurement of specific types of values (Box 5.1).

These measures of value can be incorporated into economic decision
support frameworks such as cost-benefit analysis (CBA) or cost-effectiveness
analysis (e.g., Boardman et al., 2006) to help evaluate potential adaptation
or mitigation responses. When using a CBA to compare costs and benefits
of projects or policies with long-term effects, it is common practice to
reduce, or discount, future costs and benefits. This is particularly relevant
and problematic for ocean acidification because outcomes much further in
the future than are typical of economic analysis will need to be considered.

BOX 5.1
Quantifying the Net Benefits Associated
with Non-market Goods

There are two major types of non-market values: use values and non-use values. Use values are related to observable human use (though not necessarily consumption) of a resource. Examples in the marine environment include recreational use such as beach use or scuba diving to view ocean life. Non-use values are those not related to present or future use. Examples include the value people place on the continued existence of something (existence value) or on ensuring the continued existence or availability of something for future generations (bequest value).

There are a number of common methods to quantify use values. Revealed preference methods—observing and analyzing actual human behavior—can be used to measure certain types of use values; for example, by studying the choices people make about recreation. Defensive behavior methods can also approximate non-market use values based on analysis of expenditures to avoid or mitigate environmental damage; for example, the costs associated with building groins or sea walls to prevent property damage that might otherwise have been prevented by salt marshes. However, the costs of avoiding or mitigating losses do not necessarily equate with the value of what is or would be lost, so care should be taken in using these methods to quantify value. Stated-preference methods that utilize surveys can by used to estimate non-use values. Stated preference methods such as choice experiments can also be used to evaluate the relative value of alternative policies or outcomes without necessarily monetizing them. Benefit-transfer methods, which transfer value estimates from studies in other locations, are among the most commonly applied methods for non-market valuation by government agencies (e.g., see U.S. EPA, 2002 and Griffiths and Wheeler, 2005).

The choice of discount rate in such analyses is thus likely to be both critical to the valuation and highly controversial (Box 5.2).

There is considerable uncertainty regarding the potential impacts of ocean acidification and how those impacts might be mitigated or changed by future human actions. When outcomes from different courses of action are uncertain but the probabilities of discrete alternatives occurring can be quantified, economists often apply an *expected value* or *expected utility* framework to provide a single measure of value that can be compared with the value of some other course of action (Box 5.3).

There are a number of methods beyond those outlined in Box 5.3, such as using expert panels or multi-attribute utility theory (Kim et al., 1998), that can be used to assist in determining appropriate investments in acidification research and devising policies. Each of these methods has strengths and weaknesses, and care must be taken to choose the most

BOX 5.2
Discounting

Cost-benefit analysis of policies or projects that involve costs and benefits occurring over an extended period of time will generally apply a discount rate to both future benefit and future costs. Discounting reflects the actual preferences of people for earlier consumption or delayed costs, as well as the expected growth in real consumption for future generations (Ramsey, 1928); it is generally accepted as a means of aggregating benefits and costs over time. In private investment decisions the discount rate may reflect the opportunity cost of capital. However, discounting may lead to unintended consequences when used to assess outcomes over very long time horizons. For example, in a cost-benefit analysis of a program designed to avoid a loss of $100 billion one hundred years in the future, it would be worth spending up to $24.7 billion on that program today using a discount rate of 1.4%. However, applying a discount rate of 6% would suggest it is only worth spending $247 million today on that program. Consequently, the choice of a discount rate can be extremely important in analyses of decisions with very long term implications, and can greatly alter how policies are designed and ranked (e.g., see reviews of Stern [2006] by Nordhaus [2007] and Weitzman [2007]). The discount rate is particularly critical when evaluating actions that may require large up-front costs to forestall undesirable outcomes far in the future. Some economists have proposed using low discount rates (e.g., Stern, 2007) or alternative discounting approaches for projects with long-duration effects (see Boardman et al., 2006 for a discussion of these). However, there is a lack of consensus on what discount rates or approaches should be used to evaluate decisions and design policies that will impact future generations. Therefore, it may be desirable to present policy makers with estimates of net present value reflecting alternative discount rates so that the sensitivity of the result to the discount rate is clear.

appropriate method for the assistance required and the available data. It is also important to note that performing long-time frame analysis presents difficulties for all of these analysis methods because of the challenges in weighting changes that occur far in the future.

There are a variety of important factors that determine how easily and how quickly (human) communities may cope with and adjust to the impacts of ocean acidification. These include the formal and informal institutions that determine how responses are carried out, the education and training of the affected individuals, cultural values, and alternative employment availability (Kelly and Adger, 2000; Adger, 2003; Tuler et al., 2008).[2]

---

[2] The range of issues and research questions associated with vulnerability and adaptation is broad. Though their focus is on climate change, the compiled papers in Adger et al. (2009) cover many of the issues that may be relevant to vulnerability and adaptation to ocean acidification.

BOX 5.3
Decision Making Under Uncertainty

Expected value is simply a weighted average of the values of the potential alternative outcomes where the weights represent the probabilities that certain states of
nature will occur. For example, if a particular policy has a 20% chance of providing
a benefit of $120 million and an 80% probability of accomplishing nothing, and the
cost of the policy is $20 million (e.g., a net benefit of –$20 million) the expected
value of the policy is $4 million (0.2*(120–20) + 0.8*(0–20) In an *expected value*
framework bad outcomes are not given more weight than good ones, but an
expected utility framework may weight losses more heavily than gains reflecting
risk aversion. Additional value or alternative decision criteria should be considered
in evaluating policies that prevent irreversible losses of uncertain value. The loss
of the opportunity to learn more before making a decision represents an added
cost that is called quasi-option value (Arrow and Fisher, 1974). In some cases
policy makers may choose to use a *safe minimum standard* approach. Rather than
attempt to value the loss, policies believed sufficient to ensure that the loss is not
incurred are implemented unless the costs of doing so are catastrophic. Ciriacy-
Wantrup and Phillips (1970) explained that "here the objective is not to maximize
a definite quantitative net gain but to choose premium payments and losses in
such a way that maximum possible losses are minimized." Though somewhat
flawed from an economic logic and philosophical perspective, the safe minimum
standard approach is reflected in numerous policies, including the Endangered
Species Act.

Access to capital or other resources is also likely to be important. It has been noted that strategies to cope with and adapt to impacts of climate change in the short run may not necessarily facilitate proactive adaptation and enhancement of social welfare in the longer term (Dasgupta, 2003) and may even be counterproductive (Scheraga and Grambsch, 1998). For example, emergency aid that allows a fishery-dependent community to sustain itself and maintain fishing infrastructure during a fishery collapse may be counterproductive if collapses are expected to be more frequent and severe in the future. In such cases investing in developing alterative economic opportunities may be more useful. The importance of focusing on long-run adaptation may be particularly important for ocean acidification because it is a slow driver of change with long-term effects and the potential for ecological regime shifts. Notwithstanding the potential for conflicts between different adaptation strategies, a great deal of synergy may occur among actions to facilitate adaptation to ocean acidification and other changes such as climate change, both cyclic and secular. However, in light of the variability in these factors, socioeconomic analysis should not be a one-time event but an iterative process that adjusts with the identification of stakeholders and the impact of ocean acidification upon them. As research is performed and the effects of ocean acidification are better defined, the results of the socioeconomic analysis may change, and as a result, the research needs and adaptation policies may also need to be adjusted.

It may be nearly impossible to predict how acidification will affect some ecosystem services. Indeed, the objective of prediction itself may, by necessity, be set aside for something far less ambitious—such as general understanding of basic trends or improved appreciation of risks and thresholds. Since many impacts may be hard to predict with accuracy, the development of adaptation strategies that are robust to uncertainty will be an important task for decision support (Edwards and Newman, 1982; Keeney, 1992; National Research Council, 1996; von Winterfeldt and Edwards, 1986; Kling and Sanchirico, 2009). Even when we do not fully understand the processes through which ocean acidification will effect changes in ecosystems and ecosystem services, it is useful to develop models to test the implications of alternative plausible hypotheses to provide insight into the range of possible outcomes. Sensitivity analysis can then be used to identify the assumptions and parameters of the models that most heavily impact predictions which can help target limited resources toward research aimed at the information that is likely to be of greatest value.

5.2 MARINE FISHERIES

United States wild marine fisheries had an ex-vessel value of $3.7 billion in 2007; mollusks and crustaceans comprised 49% of this commercial harvest (National Oceanic and Atmospheric Administration, 2008; Cooley and Doney, 2009). Ocean acidification may affect wild marine fisheries directly by altering the growth or survival of target species, and indirectly through changes in species' ecosystems, such as predator and prey abundance or critical habitat. This may lead to changes in abundance or size-at-age of target species, which could ultimately result in changes to sustainable harvest levels. Several experimental studies have observed the effects (positive and negative) of ocean acidification on calcification in commercially important species (e.g., Green et al., 2009; Miller et al., 2009; Ries et al., 2009; Gazeau et al., 2007). Shellfish fisheries are presumed to be particularly vulnerable to ocean acidification because of the effect on shell formation especially during early life stages (Kurihara, 2008). Many important plankton species are calcifiers, and their decline or collapse could adversely affect target species through changes in food web interactions. Fisheries could also be affected by changes in critical habitat. This could include disruption or degradation of biogenic habitat structures formed by marine calcifiers such as corals and oysters, but could also include increases in seagrass and mangrove habitats with increased CO_2 (Guinotte and Fabry, 2009). There may also be synergistic effects of increased acidification and other stressors, such as changes in water temperature associated with global climate change.

The impacts of ocean acidification on marine fisheries are likely to vary greatly over time and across species and locations, and there may be localized impacts in areas with upwelling or large freshwater input before average ocean pH falls. Studies to date have been limited to only a few commercially relevant species and have been focused on individual organisms, not on predicting the overall impacts for a target stock or species.

Ocean acidification may result in substantial losses and redistributions of economic benefits in commercial and recreational fisheries. Although fisheries make a relatively small contribution to total economic activity at a national and international level, the impacts at the local and regional level and on particular user groups could be quite important. Further, the net impact on social benefits will depend on whether adequate projections are available to allow affected fisheries to plan for change, as well as the ability of those fishery participants and communities to adapt.

The expected long lead time of acidification impacts relative to the time scales of fisheries investments makes present day valuation a challenge. For example, a snapshot of producer surplus today may substantially underestimate future producer surplus because of the likely increase in seafood demand associated with increased population and income. Rebuilding depleted fish stocks, now mandated by law, could lead to increased catches and reduced costs (Worm et al., 2009). Furthermore, many fisheries today are overcapitalized and inefficiently regulated. New "catch share" management systems being implemented in a number of U.S. fisheries provide fishermen with incentives and more flexibility to reduce harvest costs and increase the quality and value of catch (and thus net value of fisheries) as well as promote rebuilding (Worm et al., 2009; Costello et al., 2008). Taken together, these factors suggest that the potential losses from ocean acidification could be higher than projected on the basis of the current value of fisheries.

For recreational fisheries, net "consumer surplus" values must be estimated with non-market valuation techniques. As with commercial fisheries, long-term projections are likely to be highly uncertain since the number of recreational fishermen will change. Understanding likely trends in future participation in a particular fishery may help increase the accuracy of longer-term predictions.

A change in the production of a particular commercial fishery as a result of ocean acidification can be expected to result in a change of income and jobs for sellers of inputs (e.g., commercial fishing gear), processors, retailers, recreational fishing outfitters and so on. Secondary impacts such as income and job losses for sellers of inputs or fish processors are generally excluded when determining the change in net benefits, especially from a longer-term perspective, since the affected labor and capital resources can be redeployed. However, these economic impacts may be minimized and the ability of communities to adapt improved if there is good information available with sufficient lead time to allow for planned adjustments to impacts.

Beyond the value of commercial or recreational shellfish harvests, shellfish resources such as oyster reefs and mussel beds provide valuable ecosystem services. These include augmented finfish production (Grabowski and Peterson, 2007), improved water quality and clarity that can benefit submerged aquatic vegetation (Newell, 1988; Newell and Koch, 2004) and increase recreational value by improving beach and swimming use (Henderson and O'Neil, 2003). Shellfish beds can also reduce erosion of other estuarine habitats such as salt marshes by attenuating wave energy (Meyer et al., 1997; Henderson and O'Neil, 2003).

Individuals, companies, and communities involved in fisheries may be able to adapt to changes in allowable catch levels caused by ocean acidification in a variety of ways. Timely information could improve their decisions about long-term investments, including reallocation to different fisheries, diversification into multiple fisheries, or choosing a non-fishing occupation. All of these choices are strongly influenced by the culture, values, and social institutions surrounding fishing communities; therefore, adaptation responses must take these factors into consideration if they are to be effective (Coulthard, 2009). Since accurate predictions of what fisheries will be impacted when are unlikely, it is also important to identify management strategies that are robust to uncertainty and unexpected change. The potential consequences of ocean acidification may take many years to be realized but will persist for a very long time. To determine the appropriate responses to ocean acidification it is important to reduce uncertainty about when the impacts of ocean acidification will occur. Individuals and businesses involved in fisheries are likely to be interested primarily in impacts expected to occur within 20 years or less. Because of customary practices and typical discount rates applied to capital investments, projections of changes 5 to 10 years in the future are likely to be of greatest interest.

Many of the critical decisions for fisheries are made by fishery managers who must design harvest strategies and management systems. Current U.S. law requires fishery managers for federal fisheries to set reference points for biomass and exploitation rates in relation to maximum sustainable yield (MSY). These reference points, based on long time-series that reflect past conditions, will overestimate the productivity and target biomass for some species that are negatively affected by ocean acidification (i.e., lower MSY), resulting in unrealistic rebuilding requirements. The reverse may be true for other fish stocks that are positively affected by ocean acidification. In both cases, the benefits from the fishery will be reduced if reference points are not adjusted to reflect changes in a fishery's productivity. Fisheries in state waters are not subject to the Magnuson-Stevens Fisheries Conservation and Management Act (the primary U.S. law regulating marine fisheries), and guidelines on controlling overfishing or rebuilding fish stocks vary, but managers of state fisheries face the same forecasting and planning challenges as their federal counterparts.

5.3 MARINE AQUACULTURE

Since 2005, there have been many failures in oyster hatcheries along the U.S. west coast. While the cause is unknown, some attribute the failures to ocean acidification and the oyster industry has already begun to make investments in water treatment and monitoring (Welch, 2009). This underscores the urgent need for decision support for the marine aquaculture industry. It is presently unclear which aquaculture species will be impacted by ocean acidification; however, as the previous discussion of wild fisheries suggests, shellfish appear at greatest risk. Impacts on crustaceans or finfish aquaculture are presently less clear.

Many issues confronting wild fisheries also affect marine aquaculture. Estimates of the gross value of aquaculture at risk from ocean acidification (e.g., $240 million for U.S. marine aquaculture in 2006, of which $150 million was for shellfish) provide some sense of the scale of potential harm, but do not provide a measure of the net benefits that may be lost. Those can be measured through standard market-based analyses of producer and consumer surpluses (see Box 5.4) (from imported as well as domestic aquaculture) to the extent data are available. Because U.S. production has been limited mainly by markets and regulatory requirements, it is hard to forecast the level of aquaculture production a few decades from now. If aquaculture production increases significantly, the potential losses in net benefits from ocean acidification could be much higher.

Even though aquaculture faces some of the same threats as wild fisheries, the research and monitoring needs and ability to respond to threats is much different. Aquaculturists can protect against ocean acidification by changing the species or broodstock they raise, relocating operations and, in some cases, by altering seawater chemistry (e.g., in intensive culture operations and hatcheries). These decisions will require information about the probability, frequency, magnitude, and timing of potential future problems created by ocean acidification. In some cases, large investments with long payoff horizons will be at stake, so information on

---

BOX 5.4
Producer and Consumer Surplus

Gross revenues provides a rough indicator of the value of a fishery, but may not provide a good estimate of net societal benefits associated with that fishery (and thus the potential loss in value). A preferable approach is to project changes in producer and consumer surplus. Producer surplus is the difference between the revenues and the full costs associated with producing a good. Consumer surplus is the difference between what consumers pay for a good and the maximum they would be willing to pay. In addition to changes in producer and consumer surplus from U.S. fisheries, net benefits to the U.S. population could be affected by loss of consumer surplus from imported seafood. Other ecosystem services such as recreational fishing also provide consumer surplus—the value participants place on the activity itself less the expenditures they incur (e.g., travel costs, boats, fuel, gear).

---

expected impacts several years away may be useful. But, as with conventional fisheries, threats of changes 5 to 10 years in the future are likely to be of greatest interest.

5.4 TROPICAL CORAL REEFS

Coral reefs provide many valuable ecosystem services, including direct use values for recreation, e.g., diving, snorkeling, and viewing; indirect use values of coastal protection, habitat enhancement, and nursery functions for commercial and recreational fisheries; and preservation values associated with diverse natural ecosystems (Brander et al., 2007). Two coral species are listed as threatened under the U.S. Endangered Species Act—the elkhorn coral *Acropora palmata* and the staghorn coral *Acropora cervicornis*—with two others considered "species of concern" (National Ocean and Atmospheric Administration, 2009a). Tropical coral reefs also provide habitat for other protected species. According to one estimate, coral reefs are estimated to provide around $30 billion in net annual benefits globally of which some $5.7 billion is associated with fisheries, $9 billion with coastal protection, $9.6 billion with tourism and recreation, and $5.5 billion with preserving biodiversity (Cesar et al., 2003). While only about $1.1 billion is attributed to coral reefs in U.S. waters, U.S. citizens derive value from non-U.S. reefs. Many coastal populations in less developed regions of the world are dependent on reef-based fisheries for food, including people residing in U.S. territories and protectorates. Degradation or loss of reefs could undermine regional food security and have political and security implications.

The value of reefs can vary greatly and there is little consistency or agreement on methods for economic valuation. A meta-analysis of coral reef recreational valuation studies shows a wide variation of estimated values (net value of site visits) only partially explained by site charac-teristics (Brander et al., 2007). The study did find significantly higher values for reefs with larger areas, more dive sites, and fewer visitors. If the number of reefs and associated biodiversity declines over time, the value of those that remain can be expected to increase due to scarcity. Consequently, the marginal damage associated with increased reef losses would be expected to increase.

The tropical coral reef sector is somewhat different from the previous two sectors in that it represents a single ecosystem with a wider range of user groups that have different (and sometimes conflicting) values and goals. There are many potential users of information about ocean acidifi-cation impacts on tropical coral reefs, including a variety of government agencies that manage reefs (e.g., NOAA National Marine Sanctuaries Pro-gram), non-governmental conservation groups that work to protect reefs (e.g., Conservation International, World Wildlife Fund), tourism and rec-reation industry groups, native communities, and others that rely on the ecosystem services provided by reefs. Information on expected impacts on coral reefs and the vulnerabilities of these various groups may allow users to prepare for and adapt to changes. While there is virtually no information on decision support specific to ocean acidification impacts on coral reefs, there is a growing body of literature on possible management responses for the impacts of climate change (e.g., Johnson and Marshall, 2007; Keller et al., 2008; West et al., 2009). Given the similarities of the two problems, the following discussion applies the same principles toward responding to ocean acidification.

Mitigation is one possible response to predicted impacts. Analysis of the predicted impacts on coral reefs can be used to complement argu-ments to mitigate carbon dioxide emissions on a global scale. In addi-tion, small reefs having important features may warrant local mitigation actions such as using carbonates to buffer seawater, but the effectiveness and associated ecological risks have not been studied. For large-scale operations, this is unlikely to be economically feasible.

The other class of management response is to promote resilience in vulnerable components of the coral reef ecosystem and associated human communities. This will allow the system to better resist and recover from disturbances caused by acidification and is an ideal management approach given uncertainty in predictions of impacts. Approaches for managing for resilience include reducing other anthropogenic stressors such as pollution, overfishing, or habitat destruction. Managers of reef-related fisheries might need to adjust catch and effort levels to reflect reductions in productivity. Reef managers could focus protection efforts on critical elements of the reef ecosystem. For example, herbivores have been identified as a key functional group for maintenance of coral reef ecosystems; protection efforts could ensure that herbivores are afforded special protection (Johnson and Marshall, 2007). Another example is identifying and protecting refugia—areas that are less affected by ocean acidification and other stressors and that can serve as a refuge for organisms (Johnson and Marshall, 2007; West et al., 2009). It is also important to promote the social and economic resilience and adaptive capacity of users that rely on tropical coral reefs. All of this will require a great deal more information on both the biological impacts of ocean acidification on coral reefs as well as the socioeconomic systems that will be affected.

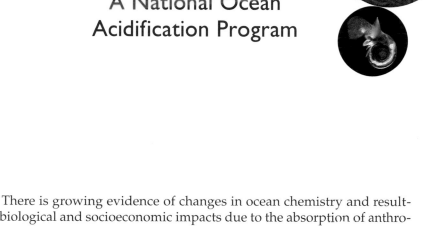

6

A National Ocean
Acidification Program

There is growing evidence of changes in ocean chemistry and resulting biological and socioeconomic impacts due to the absorption of anthropogenic CO_2 into the ocean, as summarized in chapters 2 through 5. The changes in ocean chemistry are already being detected, and because the relationship between atmospheric CO_2 and seawater carbonate chemistry is well understood, future changes can also be projected. What is less predictable is the affect these changes will have on organisms, ecosystems, and society. However, there is strong evidence that acidification will affect key biological processes—calcification and photosynthesis, for example—and that it will affect different species in different ways. This will result in ecological "winners and losers," meaning some species will do better than others in a lower pH environment, and ultimately, this will cause shifts in marine community composition and ecosystem services.

Acidification is happening globally and many ecosystems will be affected. Coral reefs appear to be particularly vulnerable because of the sensitivity of reef-builders to changes in seawater carbonate chemistry, compounded with other stressors such as climate change and overfishing. Coral reef ecosystems provide many critical resources that support a number of services, including fishing, recreation and tourism, and storm protection. They are also highly diverse ecosystems with intrinsic natural beauty whose existence alone holds high value for society. Individuals who manage coral reefs, as well as the local communities that rely on the reefs, are in urgent need of information that will allow them to mitigate and adapt to acidification impacts. Reefs are one example, but there are also many commercially-important fisheries and aquaculture species that may be vulnerable to, or may benefit from, acidification. Calcifying mollusks and crustaceans, which are important species for both aquaculture and wild harvest fisheries, and fish habitats essential for many marine species (e.g., oyster reefs, seagrass beds), are other examples. As research continues, many other sectors, communities, and decision makers that could feel an impact from acidification are likely to be identified. A better understanding of these potential biological and socioeconomic effects than we have today, as well as an ability to forecast changes, is needed for fishery managers, industry, and human communities to plan and adapt.

CONCLUSION: The chemistry of the ocean is changing at an unprecedented rate and magnitude due to anthropogenic carbon dioxide emissions; the rate of change exceeds any known to have occurred for at least the past hundreds of thousands of years. Unless anthropogenic CO_2 emissions are substantially curbed, or atmospheric CO_2 is controlled by some other means, the average pH of the ocean will continue to fall. Ocean acidification has demonstrated impacts on many marine organisms. While the ultimate consequences are still unknown, there is a risk of ecosystem changes that threaten coral reefs, fisheries, protected species, and other natural resources of value to society.

The U.S. federal government has shown a growing awareness of and response to concerns about the impacts of ocean acidification, and has taken a number of steps to begin to address the long-term implications of ocean acidification. Currently, there is no formal national program on ocean acidification; however, several federal agencies have shifted (or plan to shift) funds to ocean acidification activities (Ocean Carbon and Biogeochemistry Program, 2009a). The National Oceanic and Atmospheric Administration (NOAA) began studying the impacts of anthropogenic CO_2 on the marine carbonate system in the North Pacific in the 1980s (Feely and Chen, 1982; Feely et al., 1984, 1988) and continues to expand its research and observational efforts (e.g., Feely et al., 2008; Gledhill et al., 2008; Meseck et al., 2007). NOAA, the National Science Foundation (NSF), and the National Aeronautics and Space Administration (NASA) have also provided extramural support for workshops, planning efforts, facilities, and research (Congressional Research Service (U.S. CRS), 2009; National Science Foundation, 2009; Paula Bontempi, NASA, personal communication). In the 110th and 111th sessions, the U.S. Congress demonstrated concern over the problem of ocean acidification, holding multiple hearings and passing the Federal Ocean Acidification Research And Monitoring (FOARAM) Act of 2009 (Congressional Research Service (U.S. CRS), 2009; P.L. 111-11). The FOARAM Act of 2009 (P.L. 111-11) calls for an interagency working group (IWG) under the Joint Subcommittee on

Ocean Science and Technology (JSOST) to develop a strategic research plan and to coordinate federal ocean acidification activities.

CONCLUSION: Given that ocean acidification is an emerging field of research, the committee finds that the federal government has taken initial steps to respond to the nation's long-term needs and that the national ocean acidification program currently in development is a positive move toward coordinating these efforts.

The FOARAM Act sets out ambitious program elements in monitoring, research, modeling, technology development, and assessment and asks the IWG to develop a national program from the ground up. Fortunately, the scope of the problem is not unlike others that have faced the oceanographic and climate change communities in the past; research strategies for addressing ocean acidification can be pulled from existing programs such as the European Project on Ocean Acidification (EPOCA) and other national and multinational ocean acidification programs (see Box 6.1); other large-scale oceanographic research programs such as the Joint Global Ocean Flux Study (JGOFS); and the U.S. Global Change Research Program (USGCRP). There have also been numerous workshops and reports that have outlined recommendations for acidification research at both the international level (e.g., Raven et al., 2005; Orr et al., 2009) and within the United States (Kleypas et al., 2006; Fabry et al., 2008a; Joint et al., 2009). Fabry et al. (2008a), for example, present comprehensive research strategies for four critical major ecosystems—warmwater coral reefs, coastal margins, subtropical/tropical pelagic regions, and high latitude regions—as well as cross-cutting research issues. The U.S. reports were supported by multiple agencies (NSF, NOAA, USGS, and NASA) and represent the input of a substantial community of U.S. and international researchers. The Ocean Carbon and Biogeochemistry (OCB) Program (http://us-ocb.org/; jointly sponsored by NSF, NOAA, and NASA) has been active in supporting ocean acidification research, and produced a white paper outlining the need for a U.S. Federal Ocean Acidification Research Program (Ocean Carbon and Biogeochemistry Program, 2009a). Finally, the components of a global ocean acidification monitoring program have been proposed by a large cohort of researchers from the international oceanographic community (Feely et al., 2010). Therefore, the committee had a wealth of community-based input upon which it could base its recommendations for a National Ocean Acidification Program.

CONCLUSION: The development of a National Ocean Acidification Program will be a complex undertaking, but legislation has laid the

BOX 6.1
Existing Ocean Acidification Programs

This box briefly describes three (of several) existing national and multinational ocean acidification research programs to show some similarities and differences in program elements. It also describes one program, the IMBER/SOLAS Ocean Acidification Working Group, which is not a primary research program per se, but instead works as a coordinating body.

European Project on OCean Acidification (EPOCA): EPOCA was launched as a result of the submission of a proposal to an open call by the European Union (EU). The overall goal is to advance understanding of the biological, ecological, biogeochemical, and societal implications of ocean acidification. It is a four year program which began in June 2008. The project budget is €15.9M, with a €6.5M contribution from the EU. The project plans were developed by representatives of 10 core partners and they define a complete project with goals and deliverables. EPOCA brings together more than 100 researchers from 27 institutes and 9 European countries. EPOCA has several advisory panels, including a Reference User Group which works with EPOCA to define user-related issues such as the types of data and analysis that will be most useful to managers. There is also a project office that coordinates EPOCA activities.

From: http://www.epoca-project.eu/

Biological Impacts of Ocean ACIDification (BIOACID): BIOACID is a German national initiative that came as an unsolicited proposal to the German Ministry of Education and Research. The purpose of BIOACID is to assess uncertainties, risks, and thresholds related to the emerging problem of ocean acidification at molecular, cellular, organismal, population, community and ecosystem scales. Planning began in 2007, led by a 6-member group and with a bottom-up, open competition approach among all interested German institutes and universities conducting marine-oriented research. The project began in September 2009 and is scheduled for three years (with the possibility of 3 additional years). The German government will provide 8.9M for the first three years. BIOACID involves more than 100 scientists and technicians from 14 German research institutes and universities.

From: http://bioacid.ifm-geomar.de/index.htm

United Kingdom (UK) Ocean Acidification Research Programme: The UK program was launched as a result of the submission of a proposal to an open call by the Natural Environment Research Council and the Department for Environment, Food & Rural Affairs. The overall aim of the Research Programme is to provide a greater understanding of the implications of ocean acidification and its risks to ocean biogeochemistry, biodiversity and the whole Earth system. The science and implementation plans were written by an appointed 8-member team. Unlike EPOCA and BIOACID, the research will be determined through an open solicitation for individual proposals. The project will begin in mid 2010 and is scheduled for

continued

BOX 6.1 Continued

5 years with £12M funding from the UK government. The project is being managed by representatives of the UK government with input from a scientific Programme Advisory Group.

From: http://www.nerc.ac.uk/research/programmes/oceanacidification/

IMBER/SOLAS Ocean Acidification Working Group: This working group was initiated jointly between the Integrated Marine Biogeochemistry and Ecosystem Research (IMBER) and the Surface Ocean Lower Atmosphere Study (SOLAS)— two international oceanographic research programs—as a subgroup of the Ocean Carbon working group which coordinates seamless implementation of ocean carbon research between the two programs. Unlike the other programs, it is not supporting primary research but instead will coordinate international research efforts in ocean acidification and undertake synthesis activities in ocean acidification at the international level. The 9-member subgroup was launched in September 2009.

From: http://www.imber.info/C_WG_SubGroup3.html

---

foundation, and a path forward has been articulated in numerous reports that provide a strong basis for identifying future needs and priorities for understanding and responding to ocean acidification.

An ocean acidification program will be a complex undertaking for the nation. Like climate change, ocean acidification is being driven by the integrated global behavior of humans and is occurring at a global scale, but its impacts are likely to be felt at the regional and local level. It is a problem that cuts across disciplines and affects a diverse group of stakeholders. Assessment, research, and development of potential adaptation measures will require coordination at the international, national, regional, state, and local levels. It will involve many of the greater than 20 federal agencies that are engaged in ocean science and resource management. Investigating and understanding the problem will necessitate the close collaboration of ocean chemists, biologists, modelers, engineers, economists, social scientists, resource managers, and others from academic institutions, government labs and agencies, and non-governmental organizations. It will also involve two-way communication—both outreach to and input from—stakeholders interested in and affected by ocean acidification. Ultimately, a successful program will have an approach that integrates basic science with decision support. In this chapter, the committee describes some key elements of a successful program: a robust observing network, research to fulfill critical information needs, adaptability to new findings, and assessments and support to provide relevant information to decision makers, stakeholders, and the general public. Cutting across these elements are the needs for data management, facilities, training of ocean acidification researchers, and effective program planning and management.

6.1 OBSERVING NETWORK

Countless publications have noted the critical need for long-term ocean observations for a variety of reasons, including understanding the effects of climate change and acidification; they have also noted that the current systems for monitoring these changes are insufficient (e.g., Baker et al., 2007; Fabry et al., 2008a; Birdsey et al., 2009; National Research Council, 2009b). Currently, observations relevant to ocean acidification are being collected, but not in a systematic fashion. A global network of robust and sustained observations, both chemical and biological, will be necessary to establish a baseline and to detect and predict changes attributable to acidification (Feely et al., 2010). This network will require adequate and standardized measurements, both biological and chemical, as well as new methods and technologies for acquiring those measurements. It will also have to cover the major ecosystems that may be affected by ocean acidification, and specifically target environments that provide important ecosystem services that are potentially sensitive to acidification (e.g., fisheries, coral reefs). This network need not be entirely built "from scratch," and the program should leverage existing and developing observing systems. Even if anthropogenic CO_2 emissions remained constant at today's levels, the average pH of the ocean would continue to decrease for some period of time, and research in the area would benefit from continuous time-series data. Thus the program should consider mechanisms to sustain the long-term continuity of the observational network.

6.1.1 Measurements

The first step in developing an ocean acidification observing network is determining the requirements for biological and chemical measurements, as well as standards to ensure data quality and continuity. For ocean acidification, requirements for seawater carbonate chemistry measurements are well defined and include temperature, salinity, oxygen, nutrients critical to primary production, and at least two of the following four carbon parameters: dissolved inorganic carbon, pCO_2, total alkalinity, and pH. Methods used for these measurements are well established (Dickson et al., 2007; Ocean Carbon and Biogeochemistry Program, 2009b; Riebesell et al., 2010; see Chapter 2 of this report). As discussed in previous chapters, these values vary with depth and environment, and surface measurements alone will not provide a complete picture of conditions within the ocean. Measurements of chemical parameters should be made in different zones of interest, such as the photic zone, the oxygen minimum zone, and in deeper waters.

Unlike the chemical parameters, there are no agreed upon metrics for biological variables. In part, this is because the field is young and in part it is because the biological effects of ocean acidification, from the cellular to the ecosystem level, are very complex. While biological indicators specific to ocean acidification have not yet been defined, however, biological monitoring programs that serve a variety of applications could also be used to track responses to ocean acidification, and it would be beneficial to monitor general indicators of marine ecosystem processes to create a time series data set that will be informative to future efforts to identify correlations and trends between the chemical and biological data.

There are many potential measurements for understanding the biological response of marine ecosystems to acidification, and their relative importance will vary by ecosystem function and region. Some possible measurements include:

- rates of calcification, calcium carbonate dissolution, carbon and nitrogen fixation, oxygen production, and primary productivity,
- biological species composition, abundance, and biomass in protected and unprotected areas (Fabry et al., 2008a; Feely et al., 2010),
- the relative abundance of various taxa of phytoplankton (i.e., diatoms, dinoflagellates, coccolithophores),
- and settlement rates of sessile calcareous invertebrates (possibly commercially important species such as mussels and oysters).

Although at present we cannot predict which indicators will be informative for ocean acidification specifically, general indicators of changes in ocean and coastal ecosystems will have value for understanding changes that are a consequence of ocean acidification or other long term stressors, such as temperature. Monitoring of ecological parameters may also help researchers identify those species most vulnerable to ongoing environmental changes, including ocean acidification. As critical biological indicators and metrics are identified, the Program will need to incorporate those measurements into the research plan, and thus, adaptability in response to developments in the field should be a critical element of the monitoring program.

Resolution of the effects of ocean acidification on individuals, populations, and communities will require well-controlled manipulative experiments to assess their sensitivity and elucidate the underlying physiological mechanisms. Studies designed to understand which fundamental metabolic processes are affected by higher CO_2 or lower pH are critical to clarifying which effects on marine populations are due to ocean acidification and which to long-term or acute environmental stressors. It should also be noted that to create a time series data set that is informative for efforts to identify correlations and trends between the chemical and biological data, chemical data must be collected whenever biological data are collected. Though chemical data may stand alone, understanding the effect of ocean acidification on biological species will require that both types of data are available for analysis. Additionally, as ocean acidification is expected to be a concern into the future, data collected today will likely be analyzed by many different researchers from different areas of expertise. To facilitate archiving and sharing of information between investigators and across disciplines, the Program should support the development of standards and calibration methods for both chemical and biological samples.

Investments in technology development could greatly improve the ability to routinely measure key chemical and biological parameters in the field with expanded temporal and spatial coverage. For ocean carbonate chemistry, current instrumentation for automated pCO_2 measurements (using equilibrators and infrared detection) are robust, but similar instrumentation for continuous automated measurements of a second carbon parameter are also needed. Additional autonomous sensors could be developed for measuring particulate inorganic carbon (PIC) and particulate organic carbon (POC). There are also promising new technologies being developed for in situ pH measurements (e.g., autonomous spectrophotometric pH sensors, Seidel et al., 2008; solid state pH-sensing ion-selective field-effect transistor electrodes, Martz et al., 2008; basin-scale spatially averaged acoustic pH measurements, Duda, 2009). In the absence of direct synoptic measurements for carbonate chemistry characterization, proxy measurements have proven useful. For example, salinity and temperature have been successfully used to estimate global (Lee et al., 2006) and regional (Gledhill et al., 2008) alkalinity fields. Synoptic remotely sensed sea surface temperature measurements are available and complementary sea surface salinity measurements (SSS) should soon be available through NASA's Aquarius mission and will allow for a better understanding of current temporal and spatial variability in ocean carbonate chemistry. The temperature/salinity/alkalinity relationship may however drift in the mid- to long-term in response to acidification; sustained large-scale alkalinity measurements will therefore be needed to ground-truth proxy methods if they are to be used in the long-term. Other bio-optical sensors for in situ and remote sensing may also provide useful ocean acidification measurements. In addition, automated sensors for detecting biological parameters will need to be developed, including imaging and molecular biology tools, for detecting shifts in communities, both benthic and pelagic and across key marine ecosystems, and physiological stress markers of ocean acidification, including molecular biology tools, for key functional groups and economically important species (Byrne et al., 2010b; Feely et al., 2010). Finally, it will be important not only to develop new sensors, but also methods of deploying these on moorings, drifters, floats, gliders and underway systems.

CONCLUSION: The chemical parameters that should be measured as part of an ocean acidification observational network and the methods to make those measurements are well-established.

RECOMMENDATION: The National Program should support a chemical monitoring program that includes measurements of temperature, salinity, oxygen, nutrients critical to primary production, and at least two of the following four carbon parameters: dissolved inorganic carbon, pCO_2, total alkalinity, and pH. To account for variability in these values with depth, measurements should be made not just in the surface layer, but with consideration for different depth zones of interest, such as the deep sea, the oxygen minimum zone, or in coastal areas that experience periodic or seasonal hypoxia.

CONCLUSION: Standardized, appropriate parameters for monitoring the biological effects of ocean acidification cannot be determined until more is known concerning the physiological responses and population consequences of ocean acidification across a wide range of taxa.

RECOMMENDATION: To incorporate findings from future research, the National Program should support an adaptive monitoring program to identify biological response variables specific to ocean acidification. In the meantime, measurements of general indicators of ecosystem change, such as primary productivity, should be supported as part of a program for assessing the effects of acidification. These measurements will also have value in assessing the effects of other long term environmental stressors.

RECOMMENDATION: To ensure long-term continuity of data sets across investigators, locations, and time, the National Ocean Acidification Program should support inter-calibration, standards development, and efforts to make methods of acquiring chemical and biological

data clear and consistent. The Program should support the develop-ment of satellite, ship-based, and autonomous sensors, as well as other methods and technologies, as part of a network for observing ocean acidification and its impacts. As the field advances and a consensus emerges, the Program should support the identification and standard-ization of biological parameters for monitoring ocean acidification and its effects.

6.1.2 Establishing and Sustaining the Network

A number of existing observing systems are already conducting open ocean carbon system measurements. These include existing time series sites (e.g., Hawaii Ocean Time-Series [HOT], Bermuda Atlantic Time-Series Study [BATS]) and repeat hydrographic surveys (e.g., CLIVAR/ CO_2 Repeat Hydrography Program). Some of the sites include regular biogeochemical and biological measurements; at the HOT and BATS sites; for example, vertical profiles of inorganic carbon chemistry, nutrient, and chlorophyll concentrations and the rates of biological primary production and sinking particle flux are measured approximately monthly. Addi-tional oceanic time-series sites have been proposed (e.g., OceanSITES; Send et al., 2009).

There are also several existing marine ecosystem monitoring sites within the United States that are supported by various federal agencies, including the NSF Long-Term Ecological Research (LTER) program and NOAA National Marine Sanctuaries (Table 6.1). Monitoring is also con-ducted within the National Estuarine Research Reserve System under a partnership between NOAA and the coastal states. In addition, EPA is mandated to conduct monitoring within certain sanctuaries (e.g., the Florida Keys Marine Sanctuary), and conducts the Environmental Moni-toring and Assessment Program (EMAP). There also exist formal and informal networks of coastal marine laboratories that provide opportu-nities for assessing past historical conditions and trends, leveraging on-going observation programs, and establishing new observational systems and process studies.

There are two additional ocean observing systems in development within the United States: the Ocean Observatories Initiative (OOI) and the Integrated Ocean Observing System (IOOS). The NSF-supported OOI will provide a framework for sustained observations at four open-ocean sites in the north and south Atlantic and Pacific, a regional observing network off the Pacific Northwest, and a coastal pioneer array, initially to be deployed at the shelf-break off New England (Consortium for Ocean Leadership, 2009). The IOOS, a federal, regional, and private-sector part-nership, provides potential observational opportunities through a substantial network of open-ocean, coastal, and Great Lakes measurement sites and moorings (Integrated Ocean Observing System, 2009).

Many of these existing chemical and ecological monitoring sites could serve as a backbone for an ocean acidification observational network. However, to understand and manage effects of acidification fully, new observational efforts likely will be required in additional locations, in particular for ecosystems that may be sensitive to acidification but are currently undersampled. Fabry et al. (2008a) identify four broad eco-system areas that will require observations: warm-water coral reefs, subtropical/tropical pelagic regions, high latitude regions, and coastal margins. Within coastal regions, they highlight several specific areas: the Gulf of Alaska, western North American continental shelf, Bering Sea, Chukchi Sea, Arctic Shelf, the Scotian Shelf, Pacific coast of Central America, and the Gulf of Mexico.

While existing and developing observing networks obtain measure-ments relevant to ocean acidification, they were not originally designed with ocean acidification in mind and thus do not have adequate cover-age of these regions. The ocean inorganic carbon observing network is primarily in the open ocean with a U.S. coastal system just being devel-oped (Doney et al., 2004; Borges et al., 2009); in contrast, the ecological monitoring networks are almost entirely in coastal areas (see Table 6.1). Similarly, not all sites have adequate measurements of biological or chem-ical parameters relevant to ocean acidification. Current oceanic inorganic carbon monitoring programs do not always measure enough parameters to fully constrain the seawater carbonate system; additional inorganic carbon measurements could greatly increase the value of existing moni-toring programs for understanding acidification (Ocean Carbon and Bio-geochemistry Program, 2009b; Feely et al., 2010). Ecosystem monitoring sites measure a number of biological parameters, but have not yet been addressing acidification effects directly. The observing network can be further expanded into additional poorly sampled, but critical, coastal, estuarine and coral reef ecosystems by incorporating ocean acidification related measurements into existing long-term ecological monitoring stud-ies (e.g., marine Long-Term Ecological Research Network sites, NOAA Marine Sanctuaries, the National Estuarine Research Reserve System). Some systems may require finer spatial and temporal resolution of obser-vations to match the environmental variability in chemical and biological parameters (e.g., tropical coral reefs and estuaries). Fine-scale measure-ments may also be necessary and cost-effective in areas where critical ser-vices may be affected, for example in locales with intensive aquaculture.

The national ocean acidification network could also become a com-ponent of or partner with OOI and IOOS; this would allow the acidifica-tion network to leverage the assets of a developing integrated network

TABLE 6.1 Examples of Existing Federal Marine Ecosystem Monitoring Efforts that Could Be Leveraged for Ocean Acidification Observing and Research

Program Name	Location
Long Term Ecological Research Stations (NSF)	
California Current Ecosystem	California
Florida Coastal Everglades	Florida
Georgia Coastal Ecosystems	Georgia
Moorea Coral Reef	French Polynesia
Palmer Stations	Antarctica
Plum Island Ecosystems	Massachusetts
Santa Barbara Coastal	California
Virginia Coast Reserve	Virginia
National Marine Sanctuaries (NOAA)	
Channel Islands	California
Cordell Bank	California
Florida Keys	Florida
Flower Garden Banks	Texas
Gray's Reef	Georgia
Gulf of the Farallones	California
Northwestern Hawaiian Islands	Hawaii
Monitor	North Carolina (ship wreck)
Monterey Bay	California
Olympic Coast	Washington
Hawaiian Islands Humpback Whale	Hawaii
Fagatele Bay	American Samoa
Stellwagen Bank	Massachusetts
Thunder Bay	Great Lakes
National Monuments (FWS & NOAA)	
Papahānaumokuākea	NW Hawaiian Islands
Rose Atoll	American Samoa
Pacific Islands	Baker, Howland, Jarvis, Johnston, Kingman, Palmyra, and Wake Is.
Mariana Trench	Northern Mariana Islands

of observing systems. The OOI and IOOS networks complement existing U.S. subtropical ocean biogeochemical time-series stations by expanding into temperate and subpolar open-ocean environments and coastal waters, ecosystems that are currently identified as undersampled in community assessments of ocean carbon cycle and acidification research needs (e.g., Doney et al., 2004; Fabry et al., 2008a).

Thus the existing network of ocean carbon and marine ecosystem observing sites and surveys, complemented by the ongoing development of OOI and IOOS, will serve as a strong foundation upon which to build an ocean acidification observing network. However, the current network would be enhanced by adding monitoring sites and chemical and biological surveys in undersampled areas, particularly in areas of high variability (e.g., coastal regions), ecosystems projected to be vulnerable to ocean acidification (e.g., coral reefs and polar regions), and at depth. A community-based plan has been developed for an international ocean acidification observational network (Feely et al., 2010). The plan contains details on measurement requirements, information on data management, and an inventory of existing and planned monitoring sites and surveys. This document could serve as the basis for a national observing strategy.

CONCLUSION: The existing observing networks are inadequate for the task of monitoring ocean acidification and its effects. However, these networks can be used as the backbone of a broader monitoring network.

RECOMMENDATION: The National Ocean Acidification Program should review existing and emergent observing networks to identify existing measurements, chemical and biological, that could become part of a comprehensive ocean acidification observing network and to identify any critical spatial or temporal gaps in the current capacity to monitor ocean acidification. The Program should work to fill these gaps by:

- **ensuring that existing coastal and oceanic carbon observing sites adequately measure the seawater carbonate system and a range of biological parameters;**
- **identifying and leveraging other long-term ocean monitoring programs by adding relevant chemical and biological measurements at existing and new sites;**
- **adding additional time-series sites, repeat transects, and in situ sensors in key areas that are currently undersampled. These should be prioritized based on ecological and societal vulnerabilities.**
- **deploying and field testing new remote sensing and in situ technologies for observing ocean acidification and its impacts; and**
- **supporting the development and application of new data analysis and modeling techniques for integrating satellite, ship-based, and in situ observations.**

Sustainability of long-term observations is a perpetual challenge (e.g., Baker et al., 2007). Given the gradual and long-term pressure of ocean acidification on marine ecosystems, it is important to ensure continuity of an ocean acidification observing system for a decade or more, beyond the typical time period of many research grants. Lack of sustained funding models for ecological time-series is a significant issue (Ducklow et al., 2009), and innovative funding approaches will be necessary to ensure the sustained operations of the ocean acidification observational network. To be sustainable and efficient, the ocean acidification network will have to leverage, coordinate, and integrate with existing observing systems, other components of international ocean acidification observing networks, and other efforts to build national and international integrated ocean observing systems.

RECOMMENDATION: The National Ocean Acidification Program should plan for the long-term sustainability of an integrated ocean acidification observation network.

6.2 RESEARCH PRIORITIES

The previous chapters describe the current state of knowledge regarding ocean acidification and its impacts. There is not yet enough information on the biological, ecological, or socioeconomic effects of ocean acidification to adequately guide management efforts. Most of the existing research has been on understanding acute responses in a few species. Very little is known about the impacts of acidification on many ecologically or economically important organisms, their populations, and communities, the effects on a variety of physiological and biogeochemical processes, and the capacity of organisms to adapt to projected changes in ocean chemistry (Boyd et al., 2008). There is a need for research that provides a mechanistic understanding of physiological effects; estimates the lifelong consequences on growth, survival, and reproduction; elucidates the acclimation and adaptation potential of organisms; and that scales up to ecosystem-level effects taking into account the role and response of humans in those systems. There is also a need to understand these effects in light of multiple, potentially compounding, environmental stressors. For some systems, particularly corals, there is strong indication of impacts, but little information on how best to manage the affected system beyond reducing other stressors and promoting general resilience.

CONCLUSION: Present knowledge is insufficient to guide federal and state agencies in evaluating potential impacts of ocean acidification for management purposes.

The committee notes that ocean acidification research is a growing field and that there have been concerns over appropriate experimental design and techniques. For example, the interdependency of the inorganic carbon and acid-base chemistry parameters of seawater provides opportunities for multiple approaches, but also complicates the design of experiments and, in some cases, the comparison of results of different studies. This concern is expressed in the community development of *The Guide to Best Practices in Ocean Acidification Research and Data Reporting*, which provides guidance on measurements of seawater carbonate chemistry, experimental design of perturbation experiments, and measurements of CO_2 sensitive processes (Riebesell et al., 2010). The use of appropriate analytical techniques and experimental design is obviously critical. To enable comparison among studies and across organisms, habitats, and time, the use of standard protocols may be necessary.

Several recent workshops and symposia have brought together ocean acidification experts to identify critical information gaps and research priorities. In particular, detailed research recommendations on specific regions and topics exist in five community-based reports: *Ocean Acidification Due to Increasing Atmospheric Carbon Dioxide* (Raven et al., 2005), *Impacts of Ocean Acidification on Coral Reefs and Other Marine Calcifiers: A Guide for Future Research* (Kleypas et al., 2006), *Present and Future Impacts of Ocean Acidification on Marine Ecosystems and Biogeochemical Cycles* (Fabry et al., 2008a), *Research Priorities for Ocean Acidification* (Orr et al., 2009), and *Consequences of High CO_2 and Ocean Acidification for Microbes in the Global Ocean* (Joint et al., 2009). Fabry et al. (2008a) provide detailed recommendations for four critical marine ecosystems that include prioritization and timelines (immediate to long term). The committee believes this report provides adequate detail to appropriately balance short- and long-term research goals, as well as research, observations, and modeling requirements. Appendix D briefly summarizes these five reports and their overarching recommendations; the committee notes that the reports build upon each other and reflect a community consensus on research direction.

The committee surveyed these reports and compiled eight top research priorities, as well as some basic research approaches. The eight priorities are not ranked; the committee considers them complementary priorities to be addressed in parallel.

RECOMMENDATION: Federal and federally funded research on ocean acidification should focus on the following eight unranked priorities:

- **understand the processes affecting acidification in coastal waters;**
- **understand the physiological mechanisms of biological responses;**
- **assess the potential for acclimation and adaptation;**
- **investigate the response of individuals, populations, and communities;**

- understand ecosystem-level consequences;
- investigate the interactive effects of multiple stressors;
- understand the implications for biogeochemical cycles; and
- understand the socioeconomic impacts and inform decisions.

The research priorities are described below in greater detail. They are complementary to and synergistic with the observational priorities presented in Section 6.1 (Observing Network). Both elements are critical to addressing the ocean acidification questions facing the nation, and the two approaches will benefit from close integration during the planning, implementation, and synthesis phases of the program. For example, long-term time-series and coastal- and basin-scale surveys provide an essential context for short-duration field process studies; in turn, laboratory and field experiments provide invaluable mechanistic information for interpreting the temporal and spatial patterns found from observational networks (Doney et al., 2004). Because ocean acidification is an emerging scientific endeavor, the research priorities presented below cannot be expected to be as detailed or explicit as the observational priorities from Section 6.1. They form a framework of key questions that should be addressed, and the details of the experimental approaches and designs needed to solve these questions are best left to the creativity and innovation of individual researchers and research teams. Further, new priorities will undoubtedly arise over time based on new discoveries. Given the varying missions of the federal agencies that will fund and undertake acidification research, the committee has intentionally described broad priority areas derived from these reports; however, the committee encourages the agencies to refer to the reports for additional guidance.

6.2.1 Understand the Processes Affecting Acidification in Coastal Waters

Coastal margins are already subject to extreme variability in acid-base chemistry due to natural and anthropogenic inputs such as acidic discharge of river water (Salisbury et al., 2008) and atmospheric deposition of nitrogen and sulfur (Doney et al., 2007), and eutrophication of coastal waters from elevated river nutrient inputs due to land-use changes and agriculture (Borges and Gypens, 2010). However, the processes affecting the variability in coastal carbonate chemistry are presently not well understood, and better understanding of these processes will be necessary to predict and manage the responses of important organisms, ecosystems, and industries in coastal waters.

For example, the pH variability and range that a particular coastal location experiences may be strongly affected by fresh water runoff, which tends to have higher dissolved CO_2 concentrations, and hence lower pH, than ocean water. Coastal pH is also affected by variations related to the physical and chemical dynamics of the ocean water column, such as variations in upwelling intensity and source water depth. In general, deep, old waters are the most acidic ocean waters, but because they have not been in contact with the atmosphere for some time, there is little invasion of fossil fuel CO_2. However, the lifetime of waters in the thermocline of the ocean is measured in decades, so some acidification of upwelling source waters by anthropogenic CO_2 is expected and detected already, and acidification of this old water is projected to increase strongly in coming decades (Feely et al., 2008; see also Chapter 2). Field surveys and synoptic reconstructions based on satellite data are only now revealing this variability and the mechanisms driving it; additional research and observations will improve understanding of oceanographic and hydrological forcing of pH variability in coastal regions, which provide a wealth of ecosystem services and are already under tremendous stress.

Experimental research is also needed to characterize the impact of reduced carbonate ion concentrations and saturation states on non-living calcium carbonate particles, sediments, and reef structures. Laboratory and field studies indicate that the dissolution rates of unprotected carbonate materials increases sharply as the calcium carbonate saturation state drops below 1.0. In the water column, the shoaling of the saturation horizons and enhanced dissolution of sinking particles could alter the downward transport of food particles, carbon, and other materials to the subsurface ocean. In coastal environments, dissolution or weathering of carbonate sediments could partially buffer the effects of ocean acidification, but the faster dissolution rates could also lead to the reduction and eventual disappearance of reef structures that are valuable habitats.

6.2.2 Understand the Physiological Mechanisms of Biological Responses

Studies have shown effects of changes in the carbonate system on calcification, photosynthesis, carbon and nitrogen fixation, reproduction, and a range of other metabolic processes (see Chapter 3). However, the underlying mechanisms for these responses remain unclear in many cases. While data on the overall physiological responses of various organisms to acidification are useful, they are difficult to interpret and generalize without a fundamental understanding of the underlying chemical or biochemical mechanisms. An important aspect of mechanistic studies is that they may be useful in establishing fundamental critical thresholds beyond which the biochemical machinery of organisms cannot cope with the change in particular environmental parameters.

A striking example of a need for mechanistic studies is that of calcification—the biogenic formation of calcium carbonate minerals. Over the last few years, it has become clear that the apparently simple response of calcifying organisms to ocean acidification is a product of complex biochemical processes (see Chapter 3). Continued refinement of the understanding of how organisms such as coccolithophores or corals utilize carbon and precipitate carbonate minerals will improve the ability to predict organism responses and could eliminate exhaustive laboratory testing on a species by species basis.

A suite of improved genomic, molecular biological, biochemical, and physiological approaches using representative taxa are needed to better elucidate the mechanisms underlying those biological processes that show a response to ocean acidification. Particular examples of such processes, as highlighted in Chapter 3, include photosynthesis and phytoplankton carbon concentrating mechanisms, pathways for calcification, and physiological controls on acid-base chemistry. Mechanistic studies will also facilitate the development and interpretation of physiological stress markers needed as part of the observing system. But basic molecular and genetic tools are generally not available for marine organisms. Extending the genomic and proteomic data base to key species and developing new molecular tools, such as genetic transformation protocols for those species, would greatly enhance the ability to perform fundamental studies on marine organisms.

6.2.3 Assess the Potential for Acclimation and Adaptation

Acclimation is the process by which an organism adjusts to an environmental change that gives individuals the ability to tolerate some range of environmental variability. Though focused on the response of individuals to survive stress, survival of individuals can lead to population-level effects. The potential for individuals of most species to acclimate to higher CO_2 and lower pH is not known, but will become increasingly important as ocean CO_2 levels rise. Adaptation is the ability of a population to evolve over successive generations to become better suited to its habitat. Adaptation to changing ocean chemistry is likely on some level for most taxa that have sufficient genetic diversity to express a range of tolerance for ocean acidification. It remains unknown whether populations of most species possess both the genetic diversity and a sufficient population turnover rate to allow adaptation at the expected rate and magnitude of future pH/pCO_2 changes.

The persistence of various taxa under increasing ocean acidification will depend on either the capacity for acclimation (plasticity in phenotype within a generation) or adaptation (plasticity in genotype over successive generations) or a combination of both. The relative capabilities of various taxa in terms of both acclimation and adaptation will likely influence the composition of marine communities and therefore result in a range of consequences for marine ecosystems. Currently, too little is known about the ability of marine species to acclimate and adapt to ocean acidification to allow for assessments or predictions about how individuals and populations will respond over time. A greater understanding of these topics would help fill the gaps between physiological studies and population and community-level effects.

6.2.4 Investigate the Response of Individuals, Populations, and Communities

Well-controlled experiments, including perturbation experiments, observational studies (i.e., natural experiments) that exploit naturally-occurring spatial or temporal gradients or differences in ocean carbonate chemistry, and long-term observations of ecosystem responses to developing ocean acidification, are needed to investigate the sensitivity of individuals, populations, and communities to ocean acidification. Available information on the biological effects of acidification is currently limited to a few model organisms. While useful, this incomplete data set makes any forecast or assessment of possible impacts of acidification subject to error because of the strong potential for differential sensitivities to acidification among taxa. Extending experimentation to a range of representative organisms from key taxa in potentially affected ecosystems would allow for the identification of species or phyla that are particularly sensitive or insensitive to acidification. This is, of course, also the case for commercially important fish and shellfish. For aquaculture, it would be useful to identify tolerant subpopulations that may be used for selective breeding. In the case of deep sea ecosystems, about which very little is known, acquiring basic data on the effects of acidification on taxa representative of major phyla is an essential first step. There are clear indications that the sensitivity of higher organisms to acidification is often greater during early life stages. The evaluation of the sensitivity of various species to acidification must thus include due consideration of the life histories of these species. Experiments designed to detect effects on multiple life stages and on adaptation and acclimation potential are thus essential.

The present knowledge of pH and pCO_2 sensitivities of marine organisms is based almost entirely on short-term perturbation experiments. Interpreting and extrapolating such data requires a careful consideration of time scales. For example, the long-term success of organisms depends equally on their ability to overcome non-productive periods, such as seasonal low light or low nutrient periods, as on rapid growth during productive periods. As indicated by some experiments, the sensitivities of organisms to acidification may depend on the duration of exposure. Some organisms may be able to withstand the stress for only short exposure times; others may be able to acclimate or the population may adapt over long periods. In addition, measured positive and negative effects of ocean acidification on specific physiological processes may not always result in a net lifelong benefit or harm for the individual. There is thus a need to design and carry out acidification experiments that test the effect of exposure time and consider cumulative effects over the entire lifespan of an organism.

Therefore, manipulative experiments are required on a variety of scales, from laboratory culture incubations of single species to mesocosms and in situ perturbations with natural assemblages. Where feasible, it will be important to expand classical dose-response studies to encompass long-term and multi-generational high-CO_2 exposure experiments. It will also be necessary to design these studies to allow for reproduction and genetic recombination to test for (1) acclimation and adaptation potential and (2) cross-generational effects (those emerging in offspring generations). The use of paleo analogs and the improvement of paleo proxies may help to cover evolutionary timescales of longer-lived organisms.

6.2.5 Understand Ecosystem-level Consequences

There is little information on how the effects of ocean acidification on individual species will cascade through food webs, ultimately affecting the structure and function of ecosystems. Possible mechanisms for the transmission of the effects of ocean acidification through ecosystems include changes in microbial processes, nutrient recycling, species competition, species symbioses, calcium carbonate production, diseases, and others. In some cases, effects can be transmitted from remote locations. For example, a change in upper ocean productivity and plankton composition could affect deep-sea organisms through a change in the downward flux of organic matter even before the deep sea experiences acidification. Particularly difficult is the problem of predicting possible regime shifts (e.g., the collapse of a fishery or the shift from a coral-dominated to an algal-dominated system) which result from poorly understood nonlinearities in the internal dynamics of ecosystems. Future research on observations that will allow detection of indicators of regime shifts could help managers to anticipate shifts before they occur (de Young et al., 2008; Scheffer et al., 2009) and take action to either avoid them or cope with them.

Because resilience allows ecosystems to resist change, another important research challenge is how to maintain or increase resilience in marine ecosystems despite continued ocean acidification, occurring alongside increases in temperature and other stressors. To promote resilience in ecosystems threatened by ocean acidification, it will be important to understand what, when, and how keystone species or key functional groups will be affected. Ocean acidification will not only cause declines in some species, but increases in others; ways to understand the effects of both of these shifts need to be considered in future research strategies.

A suite of complementary approaches at various scales are needed to better understand and perhaps even predict ecosystem responses to acidification. These include controlled laboratory experiments on single organisms or cultures, bottle incubation microcosm experiments with natural microbial communities, mesocosm experiments in large enclosures, studies of natural high CO_2 environments, and field surveys along gradients in carbonate chemistry. In addition, modeling studies can be used to integrate our knowledge of physical, chemical, and biological processes to large scales. As illustrated in Figure 6.1, all these approaches have their advantages and inherent limitations. Whereas small-scale incubation experiments, also known as culture experiments, are well controlled and allow for high replication, they lack trophic complexity and reality. At the other extreme, in situ mesocosm and open water experiments allow for trophic complexity, but they are still limited in their spatial and temporal scales, allow for only a small number of replicates, and provide limited control of environmental conditions. Studies along natural, temporal, and spatial CO_2 gradients and in systems with high CO_2 variability, such as natural CO_2 vents, upwelling systems, coastal waters, and poorly buffered seas can provide the basis to help infer the response of marine ecosystems to future ocean acidification. These studies have the advantage of covering the "real" world, but they rarely approximate the actual ecosystems of interest and the data interpretation is often confounded by other variables. The insight gained from modeling studies is currently limited by imperfect knowledge of processes and parameters that are included in the models. To supplement these approaches, it might be possible in some cases to adapt to particular ocean ecosystems such as coral reefs the whole ecosystem manipulation approach that has been used extensively in terrestrial systems, particularly in lakes. In addition to examining the effects of ocean acidification, ecosystem studies can be designed to assess the efficacy and environmental consequences of ocean carbon management approaches including ocean acidification mitigation. Finally, insight into possible thresholds and tipping points may come from studies of other systems that undergo regime shifts.

Progress on understanding the future consequences of ocean acidification for marine ecosystems will require innovative methods for laboratory and ocean research and observation. Because studies of whole ecosystems are technically difficult, particularly in ocean settings, these

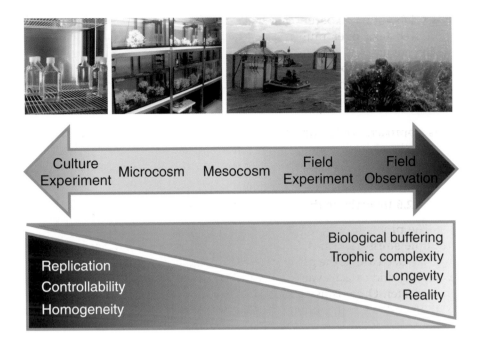

FIGURE 6.1 Experimental approaches with indication of their respective strengths and weaknesses. Photographs at top show phytoplankton bottle experiments in a culture chamber (left, courtesy of Kai Schulz, IFM-GEOMAR), cold-water corals in experimental aquaria (center left, courtesy of Armin Form, IFM-GEOMAR), an offshore mesocosm experiment in the Baltic Sea in spring 2009 (center right, Ulf Riebesell, IFM-GEOMAR), and a natural CO_2 venting site off Naples in the Mediterranean Sea (right, R. Rodolfo-Metalpa, reprinted with permission from Macmillan Publishers Ltd., Riebesell, 2008, Nature). (Gattuso et al., 2009)

types of studies will require coordination during planning and execution, perhaps including a 'task force' approach for target ecosystems. For example, research on an important and potentially vulnerable fishery (e.g., cod, salmon, and sardine/anchovy) may benefit from a coordinated research program including elements such as:

- overlap with the regional ocean acidification observation network;
- field studies documenting changes in ecosystem structure and function over natural pH gradients;
- mesocosm experiments to understand the response of phytoplankton and micrograzer communities to ocean acidification;

- laboratory experiments on the performance and survival of key food web taxa over multiple life history stages in response to ocean acidification;
- field and laboratory studies of the effects of ocean acidification on early life history phases and adults of the target fishery species; and
- whole ecosystem manipulation studies (if possible).

This approach could increase the value of focused experimental and observational studies and may be a key approach in understanding critical links in ecosystem function that are sensitive to ocean acidification.

6.2.6 Investigate the Interactive Effects of Multiple Stressors

The problem of ocean acidification is intrinsically one that involves multiple stressors (Miles, 2009). First, the increase in CO_2 concentration and the decrease in the pH and carbonate ion concentration occur simultaneously and cause a variety of other chemical changes in the chemistry of seawater. Organisms subjected to ocean acidification must also cope with the other effects of increasing atmospheric CO_2 on the climate, such as warming and increased stratification of surface waters. And, of course, marine ecosystems are affected by a variety of human activities such as fishing or pollution of coastal waters.

It is inherently difficult to study the interaction of ocean acidification with other stressors such as warming or expanding hypoxia on marine ecosystems, if only because of the large number of parameter combinations that need to be studied. In addition, environmental stresses often act synergistically, as illustrated by the simultaneous effects of high temperature events and acidification on reef building corals, or acidification and hypoxia on deep-sea crabs. For the same reason, it may also be difficult to assign any future changes in the ocean biota to a particular cause such as a decrease in pH or a decrease in carbonate ion concentration, but it will also be important to understand how acidification will impact organisms and ecosystems in light of these multiple stressors.

The perplexing problem of multiple stressors will require demanding and perhaps innovative experimental designs. In addition to factorial experiments, carefully constructed cross-site comparisons, fundamental studies of mechanisms, and synthetic modeling efforts may prove valuable. As a whole, the field would benefit from the development and discussion of unifying concepts as foundations for research on stressors that could encompass a range of efforts, from the molecular to the ecosystem level. Such a conceptual base would enable identification of similarities and differences across taxa which would be of value to the field.

6.2.7 Understand the Implications for Biogeochemical Cycles

Changes in ocean chemistry and biology due to ocean acidification have the potential to alter the oceanic cycles of carbon, nitrogen, oxygen, trace metals, other elements, and trace gases. Many of the biogeochemical priorities identified in community research plans can be grouped broadly into several interrelated themes. Ocean acidification will likely affect ocean CO_2 storage, though magnitude of the perturbation is not known because of possible counter-balancing effects. Reduced water-column and benthic calcification and faster sub-surface calcium carbonate dissolution will result in increases in surface ocean alkalinity, which should in turn enhance oceanic uptake of atmospheric CO_2. CO_2 storage is also influenced by biological export production, which may decline in some locations due to shifts away from calcifying plankton and thus reduced ballast material for sinking particles. On the other hand, export production may grow in other locations from elevated nitrogen fixation and possibly higher carbon to nutrient elemental ratios for biologically produced particulate material. These same processes would also significantly alter the subsurface distribution and cycling of carbon, nutrients and oxygen. In particular, it has been argued that elevated carbon to nutrient ratios in sinking particles could drive an expansion of tropical and subtropical oxygen minimum zones and increase marine denitrification. Ocean acidification could also influence climate and atmospheric chemistry via altered marine trace gas emissions (e.g., nitrous oxide, dimethylsulfide, and methyl halides). Finally, the impact of reduced pH on trace metal bioavailability and the chemical reactivity of dissolved organic matter are poorly understood at present.

More research is needed to understand the mechanisms governing these biogeochemical impacts and the magnitudes of the overall effects. Observations of natural systems and manipulative experiments in laboratory and field settings are essential approaches for understanding the effects of ocean acidification on biogeochemical cycles. Numerical models also provide an important tool for quantifying impacts on regional and global scales, exploring interactions among different chemical, physical and biological processes, testing hypothesis, developing projections of future behavior, and exploring feedbacks between ocean dynamics and the larger Earth system and climate. An understanding of these changes could also be informed by studying the geological record of ocean acidification. New proxy measurements, such as boron isotopes, give the promise of an estimate of surface and deep ocean pH changes over time. Although not analogous, the geological record might provide some insights on the impact of ocean acidification through quantification of the marine ecological disruption of corals, the benthos, and the plankton in the ocean and shelf environments.

6.2.8 Understand the Socioeconomic Impacts and Inform Decisions

To promote effective and informed decision making, it will be critical to integrate socioeconomic research—both for and on decision support—with natural science research. Research is needed to identify socioeconomic impacts by sector and region, to predict time frames of impacts, and to understand how to increase adaptability and resilience of socioeconomic systems. This information will enable individuals, organizations, and communities to plan for and adapt to the impacts of ocean acidification. Quantifying the cost to society of ocean acidification—its effect on the economic and social value of affected marine resources—is necessary to prioritize research efforts and decide on possible mitigation or adaptation strategies. Performing these analyses will need to be an iterative process that builds on the available research and understanding of the scope of the potential impact of acidification. As more research is performed, the boundaries of the socioeconomic analyses will shift, and research priorities may need to be adjusted.

It is important to remember that standard economic methods can be applied to market goods such as seafood, but a major part of the value of the marine resources that may be affected derives from non-market goods such as recreation or ecosystem services. These will require the use of valuation methods adapted to each type of good. Because non-market valuation studies are expensive, it may be useful to use benefit transfer methods based on studies in other areas. The impact of ocean acidification is likely to last far in the future so that valuation of its economic and social cost will need to give due consideration both to the likely increase in value of some of the affected resources in the future and to the choice of appropriate discount rate.

Understanding, predicting, and valuing impacts of ocean acidification on marine ecosystems are only the first steps. Research is also needed to improve strategies and approaches for marine ecosystem management (see section 6.3). Communities in areas with affected marine resources may be highly dependent on them both for income and sustenance. There is thus a need to assess vulnerability and adaptation capabilities of these communities over different time frames. Vulnerability assessments for fishing communities are already called for as a normal input to regulatory review for fisheries management (Clay and Olson, 2008); however, they may tend to take a short term outlook as they are typically most concerned with current or imminent changes. Since many impacts may be hard to predict with any accuracy, there is also a need to develop (and test through modeling) adaptation strategies that are robust to uncertainty about what the specific impacts will be and when they will happen. Research focused on understanding the value of advance information (e.g., more accurate and earlier predictions of biological and ecological impacts on fisheries) in improving adaptation can help determine the research expenditures that are justified in providing these predictions. There may be substantial similarity or synergy between the types of impacts on fisheries and fishing communities resulting from climate change and those due to ocean acidification. Ideally, research on vulnerability and adaptation strategies will take this into account and attempt to identify adaptation strategies that address changes on a variety of time scales and minimize conflicts between short-term and long-term objectives.

6.3 ASSESSMENT AND DECISION SUPPORT

The FOARAM Act of 2009 charges the IWG with overseeing the development of impacts assessments and adaptation and mitigation strategies, and with facilitating communication and outreach with stakeholders (P.L. 111-11). In the previous chapters, the committee identified some economic sectors and geographical regions that may be impacted by ocean acidification. The committee also identified some potential stakeholder groups, including the fishing and aquaculture industries and coral reef managers (and communities and industries that rely on services provided by reefs). However, this is not an exhaustive list; as understanding of the effects of ocean acidification improves, so will identification of stakeholder groups. Given the range of potential ecological and socioeconomic impacts outlined in the previous chapters, the need for decision support is clear.

Moving from science to decision support is often a major challenge. Indeed, it has been noted that, for climate change, "discovery science and understanding of the climate system are proceeding well, but use of that knowledge to support decision making and to manage risks and opportunities of climate change is proceeding slowly" (National Research Council, 2007b). Because ocean acidification is a relatively new concern and research results are just emerging, it will be even more challenging to move from science to decision support. Nonetheless, ocean acidification is occurring now and will continue for some time, regardless of changes in carbon dioxide emissions. Resource managers will need the ability to assess and predict these impacts on ecosystems and society, develop management plans and practices that support ecosystem resilience, identify and remove barriers to effective management response, and promote flexible decision making that adapts to challenging time scales and to altered ecosystem states (West et al., 2009).

The National Research Council (2009a) describes a comprehensive framework for decision support, including six principles for effectiveness:

1. Begin with users' needs, identified through two-way communication between knowledge producers and decision makers

2. Give priority to process (e.g., two-way communication with users) over products (e.g., data, maps, projections, tools, models) to ensure that useful products are created
3. Link information producers and users
4. Build connections across disciplines and organizations
5. Seek institutional stability for longevity and effectiveness
6. Design for learning from experience, flexibility, and adaptability. (National Research Council, 2009a)

Given the limited current knowledge about impacts of ocean acidification, the first step for the National Ocean Acidification Program will be to clearly define the problem and the stakeholders (i.e., for whom is this a problem and at what time scales?), and build a process for decision support. For climate change decision support, there have been pilot programs within some federal agencies (e.g., National Integrated Drought Information System, the Environmental Protection Agency's National Center for Environmental Assessment, NOAA Regional Integrated Sciences and Assessments [RISA] and Sectoral Applications Research Program [SARP]) and there is growing interest within the federal government for developing a national climate service to further develop climate-related decision support (National Oceanic and Atmospheric Administration, 2009b). Potentially useful tools and approaches for ecosystems and fisheries are also being developed in the context of marine ecosystem-based management and marine spatial planning (e.g., McLeod and Leslie, 2009; Douvere, 2008). The National Ocean Acidification Program could leverage the expertise of these existing and developing programs. Ocean acidification decision support could even become an integrated component of other climate service or marine ecosystem-based management programs. In addition, several recent reports have been produced on effective assessments and decision support for climate change that are equally applicable to ocean acidification (e.g., National Research Council 2005a, 2007a, b, c, 2008, 2009a, b; Adger et al., 2009); in particular, the committee notes two recent NRC reports—*Analysis of Global Change Assessments: Lessons Learned* (National Research Council, 2007a) and *Informing Decisions in a Changing Climate* (National Research Council, 2009a)—which build on previous reports and provide a strong foundation for developing an assessment and decision support strategy for ocean acidification. In particular, the FOARAM Act of 2009 (P.L. 111-11) repeatedly calls for various assessments of ocean acidification impacts. A similar mandate was given for periodic climate change assessments in the Global Change Research Act (GCRA) of 1990 (P.L. 101-606). To improve its assessment process, the U.S. Climate Change Science Program asked the NRC to look at lessons learned from past global change assessments (National Research Council,

2007a). The 11 essential elements of effective assessments determined in the NRC (2007a) report could serve as useful guidance for the development of an ocean acidification assessment strategy.

RECOMMENDATION: The National Ocean Acidification Program should focus on identifying, engaging, and responding to stakeholders in its assessment and decision support process and work with existing climate service and marine ecosystem management programs to develop a broad strategy for decision support.

6.4 DATA MANAGEMENT

Data quality and access will both be integral components of a successful program. As previously discussed, appropriate experimental design and measurements are required for high-quality data. Data reporting and archiving is important to ensure that data and associated metadata (i.e., the information about where, when, and how samples were collected and analyzed, and by whom) are accessible to researchers now and in the future. In many cases, metadata are often as important as the actual data; detailed metadata is particularly essential for manipulative experiments. Similar large-scale research programs such as U.S. JGOFS, U.S. Global Ocean Ecosystems Dynamics (GLOBEC), the LTER network, and USGCRP have developed data policies that address data quality, access, and archiving to enhance the value of data collected within these programs. *The Guide to Best Practices in Ocean Acidification Research and Data Reporting* provides guidance on data reporting and usage (Riebesell et al., 2010).

The data management component of a National Ocean Acidification Program could build on lessons learned from previous ocean research programs (e.g., Glover et al., 2006). Elements of a successful program include:

- devoting sufficient resources, about 5–10% of the total cost of the program—investments include both hardware and competent staff;
- a management office established early in the program to shepherd data management even before field programs begin;
- the development of conventions for standard methods, names, and units, as well as an agreed-to list of metadata to be collected along with the data, before field programs begin;
- an agreement among investigators to share their data with each other, leading to more rapid scientific discovery (in some cases, this requires changes in the scientific culture and incentives for investigators);
- ongoing two-way interactions between the data managers and

the principal investigators to make the database a living database and improve the final data quality; and
  • linkages between data management and data synthesis.

Data rescue efforts that compile, analyze and make publicly available existing historical data that are not currently available in electronic form would be beneficial to the field. There are many existing data management offices and databases that could support ocean acidification observational and research data, including:

  • The Biological and Chemical Oceanography Data Management Office (BCO-DMO; http://www.bco-dmo.org/) is funded by the NSF Division of Ocean Sciences and manages new data from biological and chemical oceanographic investigations, as well as legacy data from U.S. JGOFS and GLOBEC.
  • Carbon Dioxide Information Analysis Center (CDIAC; http://cdiac.ornl.gov/) is supported by the Department of Energy and provides data management support for a range of climate change projects including FACE and the Ocean CO_2 Data Project.
  • The CLIVAR and Carbon Hydrographic Data Office (CCHDO; http://cchdo.ucsd.edu/index.html) is supported by NSF and serves as a repository for CTD and hydrographic data from WOCE, CLIVAR, and other oceanographic research programs.
  • The World Data Center for Marine Environmental Sciences (WDC-MARE; http://www.wdc-mare.org/) is maintained by the Alfred Wegener Institute for Polar and Marine Research (AWI) and the Center for Marine Environmental Sciences at the University of Bremen. It is a collection of data from international (primarily European) oceanographic projects including EPOCA and BIOACID.

RECOMMENDATION: The National Ocean Acidification Program should create a data management office and provide it with adequate resources. Guided by experiences from previous and current large-scale research programs and the research community, the office should develop policies to ensure data and metadata quality, access, and archiving. The Program should identify appropriate data center(s) for archiving of ocean acidification data or, if existing data centers are inadequate, the Program should create its own.

The FOARAM Act calls for an "Ocean Acidification Information Exchange to make information on ocean acidification developed through or utilized by the interagency ocean acidification program accessible through electronic means, including information which would be useful to policymakers, researchers, and other stakeholders in mitigating or adapting to the impacts of ocean acidification" (P.L. 111-11). The committee agrees that information exchange is an important priority for the program. The Information Exchange proposed by the Act would go beyond chemical and biological measurements and also include syntheses and assessments that would be accessible to and understandable by managers, policy makers, and the general public (see section 6.3). It could also act as a conduit for two-way dialogue between stakeholders and scientists to ensure that decision support products are meeting needs of the stakeholders. A "one-stop shop" of ocean acidification information would be an extremely powerful tool, but would require resources and expertise, particularly in science communication, to perform effectively.

The committee was asked to consider the appropriate balance among research, observations, modeling, and communication. While the appropriate balance of research, observing, and modeling activities will best be determined by the IWG and individual agencies relative to their missions, the committee would like to stress the importance of communication. To successfully engage stakeholders in a two-way dialogue, the National Ocean Acidification Program will require a mechanism for effectively communicating results of the research and receiving feedback and input from managers and others seeking decision support. Inadequate progress in communicating results and engaging stakeholders, largely due to the lack of a communication strategy, has been a criticism of the U.S. Climate Change Science Program (National Research Council, 2007b). It will be important that the Ocean Acidification Information Exchange avoid a similar outcome. Both the EPOCA and OCB Program have web-based approaches for communicating science information on ocean acidification to the general public, and the National Program is encouraged to build on and learn from existing efforts in its development of an Ocean Acidification Information Exchange.

RECOMMENDATION: In addition to management of research and observational data, the National Ocean Acidification Program, in establishing an Ocean Acidification Information Exchange, should provide timely research results, syntheses, and assessments that are of value to managers, policy makers, and the general public. The Program should develop a strategy and provide adequate resources for communication efforts.

6.5 FACILITIES AND HUMAN RESOURCES

Additional facilities and trained researchers will be needed to achieve the research priorities and high quality observations described in previous sections. In some instances, ocean acidification research is likely to require large community resources and facilities, including central facilities for high-quality carbonate chemistry measurements, free-ocean CO_2 experiment (FOCE)-type experimental sites, mesocosms, wet labs with well-controlled carbonate chemistry systems, facilities at natural analogue sites, and intercomparison studies to enable integration of data from different investigators. Currently, some common facilities exist but are fairly limited. Internationally, several large-scale facilities exist or are being developed, including a mesocosm facility at the Korean Ocean Research and Development Institute in Jangmok (Kim et al., 2008) and a European Union network of aquatic mesocosm facilities (http://mesoaqua.eu/): six in-shore mesocosm facilities and a mobile off-shore mesocosm system. Ocean acidification-related facilities are also being developed within the United States: Friday Harbor Laboratories of the University of Washington (James Murray, University of Washington, personal communication) is developing analytical facilities, wet-labs, and near-shore coastal mesocosms; a FOCE prototype is in development at MBARI (http://www.mbari.org/highCO2/foce/home.htm); and "natural laboratories" have been suggested at deep and shallow CO_2 vents near the Northern Marianas Islands and other hydrothermal vents (Pala, 2009). These larger facilities may be required to scale up to ecosystem-level experiments; however, it is important to note that there are trade-offs in the various types of facilities—for example, open-ocean mesocosms are a significant scale up from coastal mesocosms but are also more costly—and that a mix of facilities will be necessary to achieve the appropriate cost-effective balance of experiments.

Ocean acidification is a highly interdisciplinary growing field, which will attract graduate students, postdoctoral investigators, and principal investigators from various fields. Training opportunities to help scientists make the transition to this new field may accelerate the progress in ocean acidification research. It may also be necessary to engage researchers in fields related to management and decision support. Preliminary capacity building efforts for ocean acidification are being developed by the OCB and EPOCA programs (e.g., http://www.whoi.edu/courses/OCB-OA/).

RECOMMENDATION: As the National Ocean Acidification Program develops a research plan, the facilities and human resource needs should also be assessed. Existing community facilities available to support high-quality field- and laboratory-based carbonate chemistry measurements, well-controlled carbonate chemistry manipulations, and large-scale ecosystem manipulations and comparisons should be inventoried and gaps assessed based on research needs. An assessment should also be made of community data resources such as genome sequences for

organisms vulnerable to ocean acidification. Where facilities or data resources are lacking, the Program should support their development, which in some cases also may require additional investments in technology development. The Program should also support the development of human resources through workshops, short-courses, or other training opportunities.

6.6 PROGRAM PLANNING, STRUCTURE, AND MANAGEMENT

The committee presents ambitious priorities and goals for the National Ocean Acidification Program, which are also echoed in the FOARAM Act and many other reports. To achieve these goals, the Program will have to lay out clear strategic and implementation plans. While the ultimate details of such plans are outside the scope of this study, there are some elements that the committee believes are necessary for a successful program. In considering recommendations on program implementation, the committee took lessons learned from large-scale research projects such as the NSF LTER Network, the USGCRP, and in particular, major oceanographic programs in its analysis and recommendations for the successful implementation of a National Ocean Acidification Program. It is important to stress, however, that a National Ocean Acidification Program—which must also link the science to decision making—will have challenges beyond these largely research-oriented programs.

The challenges to improve understanding of large-scale oceanographic phenomena with global implications has led to the rise of major U.S. oceanographic programs such as Climate VARiability and Predictability (CLIVAR), Global Ocean Ecosystems Dynamics (GLOBEC), Joint Global Ocean Flux Study (JGOFS; see Box 6.2 for case study), Ocean Drilling Program (ODP), Tropical Ocean Global Atmosphere (TOGA), and World Ocean Circulation Experiment (WOCE) programs (National Research Council, 1999). These major oceanographic programs have been recognized for their important impact on the ocean sciences, achieving an understanding of large-scale phenomena not likely without such a concentrated effort; they also produced a legacy of high-quality data, new facilities and technologies, and a new generation of trained scientists (National Research Council, 1999). In 1999, the NRC reviewed the major oceanographic programs and devised a list of guidelines and recommendations for the creation and management of large-scale oceanographic programs (see Box 6.3).

The FOARAM Act calls for the IWG to develop a detailed, 10-year strategic plan for the National Ocean Acidification Program. The committee first addresses the issue of program length. The committee agrees that a clearly defined end is appropriate because it allows for the develop-

BOX 6.2
The Joint Global Ocean Flux Study: A Model for Success

The U.S. Joint Global Ocean Flux Study (JGOFS) was a multi-agency and multi-disciplinary research and monitoring program, linked to an international program, which coordinated an ambitious agenda to study the ocean carbon cycle. The U.S. JGOFS program, a component of the U.S Global Change Research Program, was launched in the late 1980s and ran until 2005. The international program, which began a few years after the U.S. program, had over 30 participating nations; it began under the auspices of the Scientific Committee on Oceanic Research (SCOR) and eventually became a core program of the International Geosphere-Biosphere Programme (IGBP). The main goal of the JGOFS program was to understand the controls on the concentrations and fluxes of carbon and associated nutrients in the ocean. Some of the accomplishments include improved understanding of the roles of physical and biological controls on carbon cycling, improved understanding of the role of the North Atlantic in the global carbon cycle, and improved modeling of oceanic carbon dioxide uptake (National Research Council, 1999). As a result of the program, ocean biogeochemistry emerged as a new field, with emphasis on quality measurements of carbon system parameters and interdisciplinary field studies of the biological, chemical, and physical processes which control the ocean carbon cycle. U.S. JGOFS was supported primarily by the U.S. National Science Foundation in collaboration with the National Oceanic and Atmospheric Administration, the National Aeronautics and Space Administration, the Department of Energy, and the Office of Naval Research.

FROM: http://www1.whoi.edu/

ment of milestones and assessment to ensure that goals are met (National Research Council, 1999). A 10-year time frame may be adequate time to achieve many of the goals set out, but based on the experience of other major research programs, the program in its entirety may need to span a longer period (possibly 15-20 years) to incorporate an adequate synthesis phase following the field and laboratory components (e.g., Doney and Ducklow, 2006). The ultimate length of the plan will have to reflect the minimum time needed to adequately address the questions posed, and will require community input. Further, a National Ocean Acidification Program will have many elements (e.g., operational elements such as decision support) that will naturally continue beyond the initial decade; it will be critical to establish a legacy program for extended ocean acidification observations, research, and management at the outset.

In applying the guidelines from the NRC review of major oceanographic programs (National Research Council, 1999) to the design of a National Ocean Acidification Program, the committee identified some

BOX 6.3
Lessons Learned from Major Oceanographic Programs

The following paraphrases the recommendations made in *Global Ocean Science: Toward an Integrated Approach* (National Research Council, 1999) that address management of major programs. These recommendations are directly relevant to the development of a National Ocean Acidification Program.

• The federal sponsors . . . should encourage and support a broad spectrum of interdisciplinary research activities, varying in size from a collaboration of a few scientists, to intermediate-size programs, to programs perhaps even larger in scope than the present major oceanographic programs.
• Major allocation decisions (for example, extramural and internal funding of agency research) should be based on wide input from the community and the basis for decisions should be set forth clearly to the scientific community.
• . . . Sponsors and organizers of any new oceanographic program should maintain the flexibility to consider a wide range of program structures before choosing one that best addresses the scientific challenge.
• During the initial planning and organization of new major oceanographic programs, an effort should be made to ensure agreement between the program's scientific objectives and the motivating hypotheses given for funding.
• The structure should encourage continuous refinement of the program.
• The overall structure of the program should be dictated by the complexity and nature of the scientific challenges it addresses. Likewise, the nature of the administrative body should reflect the size, complexity, and duration of the program.
• All programs should have well defined milestones, including a clearly defined end. An iterative assessment and evaluation of scientific objectives and funding should be undertaken in a partnership of major ocean program leadership and agency management.
• Modelers, [experimentalists,] and observationalists need to work together during all stages of program design and implementation.
• A number of different mechanisms should be implemented to facilitate communication among the ongoing major ocean programs [and other ocean acidification programs], including (but not limited to) joint annual meetings of SSC chairs and community town meetings.
• When the scale and complexity of the program warrants, an interagency project office should be established. Other mechanisms, such as memoranda of understanding (MOU), should also be used to ensure multi-agency support throughout the program's lifetime.
• . . . The program and sponsoring agencies should establish (with input from the community) priorities for moving long-time series and other observations initiated by the program into operational mode. Factors to be considered include data quality, length [i.e., duration of program], number of variables, space and time resolution, accessibility for the wider community, and relevance to established goals.
• . . . Federal sponsors and the academic community must collaborate to preserve and ensure timely access to the data sets developed as part of each program's activities.

FROM: National Research Council, 1999.

priorities for program planning, structure, and management that will help to bring about a successful program. While the strategic plan being developed by the IWG may not contain all of the details necessary, the committee believes it is critical that an implementation plan define, at a minimum:

(1) *Goals and objectives:* Clear research, observational, and operational priorities and objectives are essential to develop a National Ocean Acidification Program. Without them, meaningful program assessment is not conceivable.

(2) *Metrics for evaluation:* Without well-defined metrics tied to both goals and objectives, meaningful or effective program operation is not possible. One cannot manage without measurement. Program operation includes and requires process, outcome, and impact evaluations—all of which depend upon well-defined measurement (National Research Council, 2005b).

(3) *Mechanisms for coordination, integration, and evaluation:* Given the proposed Program's complexity, particular care and attention will be required to assure needed coordination between, integration of, and communication among the numerous, diverse program elements and entities. Mechanisms will also need to be put in place to facilitate two-way communication among research community, decision makers, and stakeholders.

(4) *Means to transition research and observation to operations:* The plan will need to anticipate and account for the transition of some research and observational program elements to operational status. The transition plans will ensure the continuity of long-term observations and research products and facilitate the establishment, where called for, of legacy elements that continue beyond the termination of the Program.

(5) *Agency roles and institutional responsibilities:* Roles and responsibilities of every federal agency participating in the Program must be carefully specified and clearly conveyed to all of those involved (Ocean Carbon and Biogeochemistry Program, 2009a). The Program could take advantage of existing and new mechanisms for interagency funding of targeted research and observational elements.

(6) *Coordination with existing and developing national and international programs:* Ocean acidification is being recognized and taken seriously in numerous countries and diverse organizations in the United States and around the world. Given the global scope of ocean acidification, special efforts are required to take advantage of and leverage joint research and observational opportunities. Coordination is also needed to avoid possible duplications of effort. In particular, there are several different types of natural linkages with:

a. ongoing large-scale ocean and climate projects in the United States such as CLIVAR and OCB, the USGCRP, OOI, and IOOS;

  b. JSOST-led efforts on the three existing near-term priorities of the
      Ocean Research Priorities Plan: Response of Coastal Ecosystems
      to Persistent Forcing and Extreme Events, Comparative Analy-
      sis of Marine Ecosystem Organization, and Sensors for Marine
      Ecosystems;
  c. other national and multi-national carbon cycle, climate change,
      and ocean acidification programs (e.g., EPOCA, BIOACID, UK
      Ocean Acidification Research Programme, IMBER, SOLAS) and in
      particular the recently formed SOLAS-IMBER ocean acidification
      working group;
  d. international scientific bodies such as the Intergovernmental
      Oceanographic Commission (IOC), the International Council for
      Science Scientific Committee on Oceanic Research (SCOR), the
      World Climate Research Programme (WCRP), the International
      Geosphere-Biosphere Programme (IGBP), the International Council
      for the Exploration of the Sea (ICES), and the North Pacific Marine
      Science Organization (PICES) that have had demonstrated success
      in planning and coordinating international oceanographic research
      programs.
(7) *Resource requirements:* Based on the Program's stated goals and objec-
    tives, realistic resources must be identified and allocated to ensure
    success. Scrupulous attention to specific program elements, including
    those devoted to program management, data management, training,
    outreach and decision support, will be necessary. Given the dynamic
    and complex character of the ocean acidification problem, the com-
    mitment of significant resources for exploratory, innovative, and high-
    risk research will also be necessary.
(8) *Community input and external review:* Progress toward achievement
    of the Program's goals and objectives can only be measured and
    weighed based on periodic, transparent, and effective assessments
    and reviews. Peer reviews for proposals and performance are criti-
    cal to keep the Program on course toward its targeted goals and
    objectives.

**RECOMMENDATION: The National Ocean Acidification Program
should create a detailed implementation plan with community input.
The plan should address (1) goals and objectives; (2) metrics for evaluation; (3) mechanisms for coordination, integration, and evaluation;
(4) means to transition research and observational elements to operational status; (5) agency roles and responsibilities; (6) coordination
with existing and developing national and international programs;
(7) resource requirements; and (8) community input and external
review.**

   If fully executed, the elements outlined in the FOARAM Act and
recommended in this report—monitoring, interdisciplinary research,

assessment and decision support, data management, facilities, training, reporting, and outreach and communication—would create a large-scale and highly complex program that will require sufficient support. These program goals are certainly on the order of, if not more ambitious than, major oceanographic programs and will require a high level of coordination that warrants a program office. This program office would not only coordinate the activities of the program, but would also serve as a central point for communicating and collaborating with outside groups such as Congress and international ocean acidification programs. Ocean acidification is a global problem that presents challenges for research, but it also presents opportunities to share resources and expertise that may be beyond the capacity of a single nation. Therefore, international collaboration is critical to the success of the Program. It will be important to coordinate with the various other national and multi-national ocean acidification programs, as well as other international ocean carbon cycle, climate change, and marine ecosystem research programs to leverage existing resources and avoid duplication of efforts.

There are many models for such an office. The IWG called for in the FOARAM Act can be an effective approach for linking research efforts across the federal government because it resides within the JSOST, which provides for the coordination of science and technology across ocean agencies; however, a mechanism for outside input from academic scientists would be required since IWG membership is limited to federal agencies. An outside scientific steering committee consisting of representatives from the community, usually principal investigators, has been used in many major oceanographic programs (e.g., U.S. JGOFS), but this group would need to represent all stakeholders and there would still need to be a mechanism for interagency coordination of resources. An approach that combines both elements may be the best for a National Ocean Acidification Program; for example, some current interagency working groups such as the Carbon Cycle IWG work closely with an external Scientific Steering Group. Many large-scale programs (e.g., U.S. CLIVAR, U.S. GCRP) also include dedicated administrative staff that can coordinate logistics, reporting requirements, integration between program elements, communication, and other program elements. A program office is likely warranted for the National Ocean Acidification Program given the large number of stakeholders, reporting requirements, and broad research portfolio that covers both basic and applied research. Adequate resources will need to be supplied to staff a program office to support the activities of the IWG, whose participants are typically drawn from program managers and federal scientists. Where possible, efficiencies in the program office could minimize overall costs and maximize funds available to support research while completing all required tasks.

RECOMMENDATION: The National Ocean Acidification Program should create a program office with the resources to ensure successful coordination and integration of all of the elements outlined in the FOARAM Act and this report.

COMPILATION OF CONCLUSIONS AND RECOMMENDATIONS

CONCLUSION: The chemistry of the ocean is changing at an unprecedented rate and magnitude due to anthropogenic carbon dioxide emissions; the rate of change exceeds any known to have occurred for at least the past hundreds of thousands of years. Unless anthropogenic CO_2 emissions are substantially curbed, or atmospheric CO_2 is controlled by some other means, the average pH of the ocean will continue to fall. Ocean acidification has demonstrated impacts on many marine organisms. While the ultimate consequences are still unknown, there is a risk of ecosystem changes that threaten coral reefs, fisheries, protected species, and other natural resources of value to society.

CONCLUSION: Given that ocean acidification is an emerging field of research, the committee finds that the federal government has taken initial steps to respond to the nation's long-term needs and that the national ocean acidification program currently in development is a positive move toward coordinating these efforts.

CONCLUSION: The development of a National Ocean Acidification Program will be a complex undertaking, but legislation has laid the foundation, and a path forward has been articulated in numerous reports that provide a strong basis for identifying future needs and priorities for understanding and responding to ocean acidification.

CONCLUSION: The chemical parameters that should be measured as part of an ocean acidification observational network and the methods to make those measurements are well established.

RECOMMENDATION: The National Program should support a chemical monitoring program that includes measurements of temperature, salinity, oxygen, nutrients critical to primary production, and at least two of the following four carbon parameters: dissolved inorganic carbon, pCO_2, total alkalinity, and pH. To account for variability in these values with depth, measurements should be made not just in the surface layer, but with consideration for different depth zones of interest, such as the deep sea, the oxygen minimum zone, or in coastal areas that experience periodic or seasonal hypoxia.

CONCLUSION: Standardized, appropriate parameters for monitoring the biological effects of ocean acidification cannot be determined until more is known concerning the physiological responses and population consequences of ocean acidification across a wide range of taxa.

RECOMMENDATION: To incorporate findings from future research, the National Program should support an adaptive monitoring program to identify biological response variables specific to ocean acidification. In the meantime, measurements of general indicators of ecosystem change, such as primary productivity, should be supported as part of a program for assessing the effects of acidification. These measurements will also have value in assessing the effects of other long-term environmental stressors.

RECOMMENDATION: To ensure long-term continuity of data sets across investigators, locations, and time, the National Ocean Acidification Program should support inter-calibration, standards development, and efforts to make methods of acquiring chemical and biological data clear and consistent. The Program should support the development of satellite, ship-based, and autonomous sensors, as well as other methods and technologies, as part of a network for observing ocean acidification and its impacts. As the field advances and a consensus emerges, the Program should support the identification and standardization of biological parameters for monitoring ocean acidification and its effects.

CONCLUSION: The existing observing networks are inadequate for the task of monitoring ocean acidification and its effects. However, these networks can be used as the backbone of a broader monitoring network.

RECOMMENDATION: The National Ocean Acidification Program should review existing and emergent observing networks to identify existing measurements, chemical and biological, that could become part of a comprehensive ocean acidification observing network and to identify any critical spatial or temporal gaps in the current capacity to monitor ocean acidification. The Program should work to fill these gaps by:

• ensuring that existing coastal and oceanic carbon observing sites adequately measure the seawater carbonate system and a range of biological parameters;
• identifying and leveraging other long-term ocean monitoring programs by adding relevant chemical and biological measurements at existing and new sites;

- adding additional time-series sites, repeat transects, and in situ sensors in key areas that are currently undersampled. These should be prioritized based on ecological and societal vulnerabilities.
- deploying and field testing new remote sensing and in situ technologies for observing ocean acidification and its impacts; and
- supporting the development and application of new data analysis and modeling techniques for integrating satellite, ship-based, and in situ observations.

RECOMMENDATION: The National Ocean Acidification Program should plan for the long-term sustainability of an integrated ocean acidification observation network.

CONCLUSION: Present knowledge is insufficient to guide federal and state agencies in evaluating potential impacts for management purposes.

RECOMMENDATION: Federal and federally funded research on ocean acidification should focus on the following eight unranked priorities:

- understand processes affecting acidification in coastal waters;
- understand the physiological mechanisms of biological responses;
- assess the potential for acclimation and adaptation;
- investigate the response of individuals, populations, and communities;
- understand ecosystem-level consequences;
- investigate the interactive effects of multiple stressors;
- understand the implications for biogeochemical cycles; and
- understand the socioeconomic impacts and inform decisions.

RECOMMENDATION: The National Ocean Acidification Program should focus on identifying, engaging, and responding to stakeholders in its assessment and decision support process and work with existing climate service and marine ecosystem management programs to develop a broad strategy for decision support.

RECOMMENDATION: The National Ocean Acidification Program should create a data management office and provide it with adequate resources. Guided by experiences from previous and current large-scale research programs and the research community, the office should develop policies to ensure data and metadata quality, access, and archiving. The Program should identify appropriate data center(s) for archiving of ocean acidification data or, if existing data centers are inadequate, the Program should create its own.

RECOMMENDATION: In addition to management of research and observational data, the National Ocean Acidification Program, in establishing an Ocean Acidification Information Exchange, should provide timely research results, syntheses, and assessments that are of value to managers, policy makers, and the general public. The Program should develop a strategy and provide adequate resources for communication efforts.

RECOMMENDATION: As the National Ocean Acidification Program develops a research plan, the facilities and human resource needs should also be assessed. Existing community facilities available to support high-quality field- and laboratory-based carbonate chemistry measurements, well-controlled carbonate chemistry manipulations, and large-scale ecosystem manipulations and comparisons should be inventoried and gaps assessed based on research needs. An assessment should also be made of community data resources such as genome sequences for organisms vulnerable to ocean acidification. Where facilities or data resources are lacking, the Program should support their development, which in some cases also may require additional investments in technology development. The Program should also support the development of human resources through workshops, short-courses, or other training opportunities.

RECOMMENDATION: The National Ocean Acidification Program should create a detailed implementation plan with community input. The plan should address (1) goals and objectives; (2) metrics for evaluation; (3) mechanisms for coordination, integration, and evaluation; (4) means to transition research and observational elements to operational status; (5) agency roles and responsibilities; (6) coordination with existing and developing national and international programs; (7) resource requirements; and (8) community input and external review.

RECOMMENDATION: The National Ocean Acidification Program should create a program office with the resources to ensure successful coordination and integration of all of the elements outlined in the FOARAM Act and this report.

References

Adger, W.N. 2003. Social capital, collective action, and adaptation to climate change. *Economic Geography* 79(4): 387-404.

Adger, W.N., I. Lorenzoni, and K.L. O'Brien (Eds). 2009. *Adapting to climate change thresholds, values, governance.* Cambridge University Press.

Accornero, A., C. Manno, K.R. Arrigo, A. Martini and S. Tucci. 2003. The vertical flux of particulate matter in the polynya of Terra Nova Bay: Part I: Chemical constituents. *Antarctic Science* 15: 119-132.

Al-Horani, F.A., S.M. Al-Moghrabi, and D. de Beer. 2003. The mechanism of calcification and its relation to photosynthesis and respiration in the scleractinian coral *Galaxea fascicularis. Marine Biology* 142: 419-426.

Albright, R., B. Mason, and C. Langdon. 2008. Effect of aragonite saturation state on settlement and post-settlement growth of *Porites astreoides* larvae. *Coral Reefs* 27: 485-490.

Alvarez-Filip, L., N.K. Dulvy, J.A. Gill, I.M. Côté, and A.R. Watkinson. 2009. Flattening of Caribbean coral reefs: Region-wide declines in architectural complexity. *Proceedings of the Royal Society B: Biological Sciences* 276: 3019-3025. published online before print June 10, 2009. [doi:10.1098/rspb.2009.0339]

Andersen, T., J. Carstensen, E. Hernandez-Garcia, and C.M. Duarte. 2009. Ecological thresholds and regime shifts: Approaches to identification. *Trends in Ecology and Evolution* 24: 49-57.

Andersson, A.J., I.B. Kuffner, F.T. Mackenzie, P.L. Jokiel, K.S. Rodgers, and A. Tan. 2009. Net loss of $CaCO_3$ from a subtropical calcifying community due to seawater acidification: mesocosm-scale experimental evidence. *Biogeosciences* 6: 1811-1823.

Andersson, A.J., F.T. Mackenzie, and L.M. Ver. 2003. Solution of shallow-water carbonates: An insignificant buffer against rising atmospheric CO_2. *Geology* 31(6): 513-516.

Andersson, A.J., N.R. Bates, and F.T. Mackenzie. 2007. Dissolution of carbonate sediments under rising pCO_2 and ocean acidification: Observations from Devil's Hole, Bermuda. *Aquatic Geochemistry* 13(3): 237-264.

Andrews, A.H., E.E. Cordes, M.M. Mahoney, K. Munk, K.H. Coale, G.M. Cailliet, and J. Heifetz. 2002. Age, growth, and radiometric age validation of a deep-sea habitat-forming gorgonian (*Primnoa resedaeformis*) from the Gulf of Alaska. *Hydrobiologia* 471: 101-110.

Anthony, K.R.N., D.I. Kline, G. Diaz-Pulido, S. Dove, and O. Hoegh-Guldberg. 2008. Ocean acidification causes bleaching and productivity loss in coral reef builders. *Proceedings of the National Academy of Sciences* 105: 17442-17446.

Armstrong, J.L., J.L. Boldt, A.D. Cross, J.H. Moss, N.D. Davis, K.W. Myers, R.V. Walker, D.A. Beauchamp, and L.J. Haldorson. 2005. Distribution, size, and interannual, seasonal and diel food habits of northern Gulf of Alaska juvenile pink salmon, *Oncorhynchus gorbuscha*. *Deep Sea Research Part II: Topical Studies in Oceanography* 52(1-2): 247-265.

Armstrong, J.L., K.W. Myers, D.A. Beauchamp, N.D. Davis, R.V. Walker, and J.L. Boldt. 2008. Interannual and spatial feeding patterns of hatchery and wild juvenile pink salmon in the Gulf of Alaska in years of low and high survival. *Transactions of the American Fisheries Society* 137: 1299-1316.

Armstrong, R.A., C. Lee, J.I. Hedges, S. Honjo, and S.G. Wakeham. 2002. A new, mechanistic model for organic carbon fluxes in the ocean based on the quantitative association of POC with ballast minerals. *Deep-Sea Research Part II: Topical Studies In Oceanography* 49: 219-236.

Arnold, K.E., H.S. Findlay, J.I. Spicer, C.L. Daniels, and D. Boothroyd. 2009. Effect of CO_2-related acidification on aspects of the larval development of the European lobster, *Homarus gammarus* (L.) *Biogeosciences Discussions* 6: 3087-3107.

Aronson, R.B., S. Thatje, A. Clarke, L.S. Peck, D.B. Blake, C.D. Wilga, and B.A. Seibel. 2007. Climate change and invasibility of the Antarctic benthos. *Annual Review of Ecology, Evolution, and Systematics* 38: 129-154.

Arrigo, K.R. 2007. Marine manipulations. *Nature* 450: 491-492.

Arrow, K.J. and A.C. Fisher. 1974. Environmental preservation, uncertainty, and irreversibility. *The Quarterly Journal of Economics* 88(2): 312-319.

Baker, D.J., R.W. Schmitt, and C. Wunsch. 2007. Endowments and new institutions for long-term observations. *Oceanography* 20(4): 10-14.

Barcelos e Ramos, J., H. Biswas, K.G. Schulz, J. LaRoche, and U. Riebesell. 2007. Effect of rising atmospheric carbon dioxide on the marine nitrogen fixer Thrichodesmium, *Global Biogeochemical Cycles* 21: GB2028.

Barry, J.P. and J.C. Drazen. 2007. Response of deep-sea scavengers to ocean acidification and the odor from a dead grenadier. *Marine Ecology Progress Series* 350: 193-207.

Barry, J.P., B.A. Seibel, J.C. Drazen, M.N. Tamburri, K.R. Buck, C. Lovera, L. Kuhnz, E.T. Peltzer, K. Osborn, P.J. Whaling, P. Walz, and P.G. Brewer. 2003. Deep-sea field experiments on the biological impacts of direct deep-sea CO_2 injection. *Proceedings of the 2nd Annual Conference on Carbon Sequestration*, May 5-8, 2003, Alexandria, VA.

Barry, J.P., K.R. Buck, C. Lovera, L. Kuhnz, and P.J. Whaling. 2005. Utility of deep sea CO_2 release experiments in understanding the biology of a high-CO_2 ocean: Effects of hypercapnia on deep sea meiofauna. *Journal of Geophysical Research* 110: CO9S12.

Barry, J.P., P.J. Whaling, and R.K. Kochevar. 2007. Growth, production, and mortality of the chemosynthetic vesicomyid bivalve *Calyptogena kilmeri* from cold seeps off central California. *Marine Ecology* 28(1): 169-182.

Bates, N.R., J.M. Mathis, and L.W. Cooper. 2009. Ocean acidification and biologically induced seasonality of carbonate mineral saturation states in the western Arctic Ocean. *Journal of Geophysical Research*. 114: C11007. [doi:10.1029/2008JC004862]

Bathmann, U., G. Fischer, P.J. Müller, and D. Gerdes. 1991. Short-term variations in particulate matter sedimentation off Kapp Norvegia, Weddell Sea, Antarctica: Relation to water mass advection, ice cover, plankton biomass and feeding activity. *Polar Biology* 11(3): 185-195.

Bellerby, R.G.J., K.G. Schulz, U. Ribesell, C. Neil, G. Nondal, T. Johannessen, and K.R. Brown. 2007. Marine ecosystem community carbon and nutrient uptake stoichiometry under varying ocean acidification during the PeECE III experiment. *Biogeoscienes Discussions* 4: 4631-4652.

Bentova S., C. Brownlee, and J. Erez. 2009. The role of seawater endocytosis in the biomineralization process in calcareous foraminifera. *Proceedings of the National Academy of Sciences* 106(51): 21500-21504. [doi10.1073pnas.0906636106]

Berner, R.A, and Z. Kothavala. 2001. GEOCARB III: A Revised Model of Atmospheric CO_2 Over Phanerozoic Time. *American Journal of Science* 301: 182–204.

Bibby, R., P. Cleall-Harding, S. Rundle, S. Widdicombe, and J. Spicer. 2007. Ocean acidification disrupts induced defences in the intertidal gastropod *Littorina littorea*. *Biology Letters* 3: 699-701.

Birdsey, R., N. Bates, M. Behrenfeld, K. Davis, S.C. Doney, R. Feely, D. Hansell, L. Heath, E. Kasischke, H. Kheshgi, B. Law, C. Lee, A.D. McGuire, P. Raymond, and C.J. Tucker. 2009. Carbon cycle observations: Gaps threaten climate mitigation policies. *EOS, Transactions American Geophysical Union* 90(34). [doi:10.1029/2009EO340005]

Blackford, J., N. Jones, R. Proctor, J. Holt, S. Widdicombe, D. Lowe, and A. Rees. 2009. An initial assessment of the potential environmental impact of CO_2 escape from marine carbon capture and storage systems. *Proceedings of the Institution of Mechanical Engineers, Part A: Journal of Power and Energy* 223(3): 269-280.

Boardman, A.E., D.H. Greenberg, A.R. Vining, and D.L. Weimer. 2006. *Cost-benefit analysis: Concepts and practice.* 3rd edition. Prentice Hall: Upper Saddle River, NJ.

Bockstael, N.E., A. M. Freeman, R.J. Kopp, P.R. Portney, and V.K. Smith. 2000. On measuring economic values for nature. *Environmental Science & Technology* 34(8):1384.

Booth, I.R. 1985. Regulation of cytoplasmic pH in bacteria. *Microbiological Reviews* 49(4): 359-378.

Borges, A.V. and N. Gypens. 2010. Carbonate chemistry in the coastal zone responds more strongly to eutrophication than to ocean acidification. *Limnology and Oceanography* 55(1): 346-353.

Borges, A.V., S.R. Alin, F.P. Chavez, P. Vlahos, K.S. Johnson, and J.T. Holt. 2009. A global sea surface carbon observing system: Inorganic and organic carbon dynamics in coastal oceans. *Ocean Observations Conference 2009 Report.* [Online]. Available: http://www.oceanobs09.net/blog/?p=624

Boyd, P.W. 2008. Ranking geo-engineering schemes. *Nature Geoscience* 1: 722-724.

Boyd, P.W. and S.C. Doney. 2003. The impact of climate change and feedback processes on the ocean carbon cycle. In Fasham, M.J.R. (Ed.). *Ocean biogeochemistry: The role of the ocean carbon cycle in global change. Global Change - The IGBP Series.* Springer: Berlin. pp. 157-193.

Boyd, P.W., T. Jickells, C.S. Law, S. Blain, E.A. Boyle, K.O. Buesseler, K.H. Coale, J.J. Cullen, H.J.W. de Baar, M. Follows, M. Harvey, C. Lancelot, M. Levasseur, N.P.J. Owens, R. Pollard, R.B. Rivkin, J. Sarmiento, V. Schoemann, V. Smetacek, S. Takeda, A. Tsuda, S. Turner, and A.J. Watson. 2007. Mesoscale iron enrichment experiments 1993-2005: Synthesis and future directions. *Science* 315 (5812): 612.

Boyd, P.W., S.C. Doney, R. Strzepek, J. Dusenberry, K. Lindsay, and I. Fung. 2008. Climate-mediated changes to mixed-layer properties in the Southern Ocean: Assessing the phytoplankton response. *Biogeosciences* 5: 847-864.

Brander, L.M., P.J.H. van Beukering, and H.S.J. Cesar. 2007. The recreational value of coral reefs: A meta-analysis. *Ecological Economics* 63: 209-218.

Briske, D.D., S.D. Fuhlendorf, and F.E. Smeins. 2005. State-and-transition models, thresholds, and rangeland health: A synthesis of ecological concepts and perspectives. *Rangeland Ecology and Management* 58(1): 1-10.

Broecker, W.S. and T. Takahashi. 1966. Calcium Carbonate Precipitation on the Bahama Banks. *Journal of Geophysical Research* 71: 1575.

Broecker, W.S., C. Langdon, T. Takahashi, and T-H. Peng. 2001. Does carbon 13 track anthropogenic CO_2 in the southern ocean? *Global Biogeochemical Cycles* 15(3): 589-596. [doi:10.1029/2000GB001350].

Burkhardt, S., I. Zondervan, and U. Riebesell. 1999. Effect of CO_2 concentration on the C:N:P ratio in marine phytoplankton: A species comparison. *Limnology and Oceanography* 44: 683-690.

Byrne, R.H., L.R Kump, and K.J Cantrell. 1988. The influence of temperature and pH on metal speciation in seawater. *Marine Chemistry* 25: 163-181.

Byrne, R.H., S. Mecking, R.A. Feely, and X. Liu. 2010a. Direct observations of basin-wide acidification of the North Pacific Ocean. Geophysical Research Letters 37: L02601. [doi:10.1029/2009GL040999]

Byrne, R.H., M.D. DeGrandpre, R.T. Short, T.R. Martz, L. Merlivat, C. McNeil, F.L. Sayles, R. Bell, and P. Fietzek. 2010b. Sensors and systems for observations of marine CO_2 system variables. In *Proceedings of OceanObs'09: Sustained Ocean Observations and Information for Society (Vol. 2)*, Venice, Italy, 21-25 September 2009, Hall, J., Harrison D.E. & Stammer, D. (Eds.). ESA Publication WPP-306.

Cairns, S.D. 2007. Deep-water corals: An overview with special reference to diversity and distribution of deep-water scleractinian corals. *Bulletin of Marine Science* 81(3): 311-322.

Caldeira, K., R. Berner, E.T. Sundquist, P.N. Pearson, and M.R. Palmer. 1999. Seawater pH and atmospheric carbon dioxide. *Science* 286(5447): 2043.

Caldeira, K. and G.H. Rau. 2000. Accelerating carbonate dissolution to sequester carbon dioxide in the ocean: Geochemical implications. *Geophysical Research Letters* 27(2): 225-228.

Caldeira, K. and M.E. Wickett. 2003. Oceanography: Anthropogenic carbon and ocean pH. *Nature* 425: 365.

Caldeira, K., and M.E. Wickett. 2005. Ocean model predictions of chemistry changes from carbon dioxide emissions to the atmosphere and ocean. *Journal of Geophysical Research* 110: C09S04. [doi:10.1029/2004JC002671]

Carpenter, S.R., D. Ludwig, and W.A. Brock. 1999. Management of eutrophication for lakes subject to potentially irreversible change. *Ecological Applications* 9: 751-771.

Carpenter, K.E., M. Abrar, G. Aeby, R.B. Aronson, S. Banks, A. Bruckner, A. Chiriboga, J. Cortés, J.C. Delbeek, L. DeVantier, G.J. Edgar, A.J. Edwards, D. Fenner, H.M. Guzmán, B.W. Hoeksema, G. Hodgson, O. Johan, W.Y. Licuanan, S.R. Livingstone, E.R. Lovell, J.A. Moore, D.O. Obura, D. Ochavillo, B.A. Polidoro, W.F. Precht, M.C. Quibilan, C. Reboton, Z.T. Richards, A.D. Rogers, J. Sanciangco, A. Sheppard, C. Sheppard, J. Smith, S. Stuart, E. Turak, J.E.N. Veron, C. Wallace, E. Weil, and E. Wood. 2008. One-third of reef-building corals face elevated extinction risk from climate change and local impacts. *Science* 321(5888): 560-563.

Carpenter, S.R. and W.A. Brock. 2006. Rising variance: A leading indicator of ecological transition. *Ecology Letters* 9: 311-318.

Cesar, H., L. Burke, and L. Pet-Soede. 2003. *The economics of worldwide coral reef degradation*. CEEC: Arhem, Netherlands.

Ciriacy-Wantrup, S.V. and W.E. Phillips. 1970. Conservation of the California Tule Elk: A socioeconomic study of a survival problem. Biological Conservation 3(1): 23-32.

Clausen, C.D. and A.A. Roth. 1975. Effect of temperature and temperature adaptation on calcification rate in hermatypic coral *Pocillopora damicornis*. *Marine Biology* 33: 93-100.

Clark, P.U., D. Archer, D. Pollard, J.D. Blum, J.A. Rial, V. Brovkin, A.C. Mix, N.G. Pisias, and M. Roy. 2006. The middle Pleistocene transition: Characteristics, mechanisms, and implications for long-term changes in atmospheric pCO_2. *Quaternary Science Reviews* 25: 3150–3184.

Clark, D., M. Lamare, and M. Barker. 2009. Response of sea urchin pluteus larvae (Echinodermata: Echinoidea) to reduced seawater pH: A comparison among a tropical, temperate, and a polar species. *Marine Biology* 156(6): 1125-1137.

Clay, P.M. and J. Olson. 2008. Defining "fishing communities": Vulnerability and the Magnuson-Stevens Fishery Conservation and Management Act. *Human Ecology* 15(2):143-160.

Cohen, A.L., D.C. McCorkle, S. de Putron, G.A. Gaetani, and K.A. Rose. 2009. Morphological and compositional changes in the skeletons of new coral recruits reared in acidified seawater. Insights into the biomineralization response to ocean acidification. *Geochemistry Geophysics Geosystems* 10: Q07005.

Collier, R., J. Dymond, S. Honjo, S. Manganini, R. Francois, and R. Dunbar. 2000. The vertical flux of biogenic and lithogenic material in the Ross Sea: Moored sediment trap observations 1996–1998. *Deep Sea Research Part II: Topical Studies in Oceanography* 47(15-16): 3491-3520.

Comeau, S., G. Gorsky, R. Jeffree, J-L. Teyssie, and J-P. Gattuso. 2009. Impact of ocean acidification on a key Arctic pelagic mollusc (*Limacina helicina*). *Biogeosciences* 6: 1877–1882.

Consortium for Ocean Leadership. 2009. *Ocean observatories initiative.* [Online]. Available: http://www.oceanleadership.org/programs-and-partnerships/ocean-observing/ooi/ [Accessed on December 2, 2009].

Cooley, S.R. and S.C. Doney. 2009. Anticipating ocean acidification's economic consequences for commercial fisheries. *Environmental Research Letters* 4: 024007. [doi:10.1088/1748-9326/4/2/024007]

Cooley, S., H.L. Kite-Powell, and S.C. Doney. 2009. Ocean acidification's potential to alter global marine ecosystem services. *Oceanography* 22(4): 172-180.

Costello, C., S.D. Gaines, and J. Lynham. 2008. Can catch shares prevent fisheries collapse?. *Science* 321(5896): 1678-1681. [DOI: 0.1126/science.1159478]

Coulthard, S. 2009. Adaptation and conflict within fisheries: Insights for living with climate change. *In* Adger, W.N., I. Lorenzoni, and K.L. O'Brien (Eds). *Adapting to climate change thresholds, values, governance.* Cambridge University Press. pp 255-268.

Crutzen, P.J. 2006. Albedo enhancement by stratospheric sulfur injections: A contribution to resolve a policy dilemma?. *Climatic Change* 77(3-4): 211-220.

Czerny, J., J. Barcelos e Ramos, and U. Riebesell. 2009. Influence of elevated CO_2 concentrations on cell division and nitrogen fixation rates in the bloom-forming cyanobacterium *Nodularia spumigena*. *Biogeosciences* 6: 1865-1875.

Dahlhoff, E.P. 2004. Biochemical indicators of stress and metabolism: Applications for marine ecological studies. *Annual Review of Physiology* 66:183-207.

Dakos, V., M. Scheffer, E.H. van Nes, V. Brovkin, V. Petoukhov, and H. Held. 2008. Slowing down as an early warning signal for abrupt climate change. *Proceedings of the National Academy of Sciences* 105: 14308-14312.

Dasgupta, P. 2003. Social capital and economic performance: Analytics. In *Foundations of social capital*. Ostrom, E. and T.K. Ahn (Eds.). 238–57. Edward Elgar: Cheltenham, U.K.

Dayton, P.K., B.J. Mordida , and F. Bacon. 1994. Polar marine communities. *American Zoologist* 34: 90-99.

De'ath, G., J.M. Lough, and K.E. Fabricius. 2009. Declining coral calcification on the Great Barrier Reef. *Science* 323: 116-119.

de Young, B., M. Barange, G. Beaugrand, R. Harris, R.I. Perry, M. Scheffer, and F. Werner. 2008. Regime shifts in marine ecosystems: Detection, prediction and management. *Trends in Ecology and Evolution* 23(7): 402-409.

D'Hondt, S. and G. Keller. 1991. Some patterns of planktoc foraminiferal assemblage turnover at the Cretaceous-Tertiary boundary. *Marine Micropaleontology* 17: 77-118.

D'Hondt, S., M.E.Q. Pilson, H. Sigurdsson, A. Hanson, and S. Carey. 1994. Surface-water acidification and extinction at the Cretaceous-Tertiary boundary. *Geology* 22: 983-986.

Dickson, A.G., C.L. Sabine, and J.R. Christian. 2007. Guide to best practices for ocean CO_2 measurements. *PICES Special Publication* 3: 191 pp.

Doney, S.C. and H.W. Ducklow. 2006. A decade of synthesis and modeling in the U.S. Joint Global Ocean Flux Study. *Deep Sea Research Part II: Topical Studies in Oceanography* 53(5-7): 451-458.

Doney, S.C., R. Anderson, J. Bishop, K. Caldeira, C. Carlson, M.-E. Carr, R. Feely, M. Hood, C. Hopkinson, R. Jahnke, D. Karl, J. Kleypas, C. Lee, R. Letelier, C. McClain, C. Sabine, J. Sarmiento, B. Stephens, and R. Weller. 2004. *Ocean carbon and climate change (OCCC): An implementation strategy for U.S. ocean carbon cycle science.* University Center for Atmospheric Research: Boulder, CO. 108pp.

Doney, S.C., N. Mahowald, I. Lima, R.A. Feely, F.T. Mackenzie, J-F. Lamarque, and P.J. Rasch. 2007. Impact of anthropogenic atmospheric nitrogen and sulfur deposition on ocean acidification and the inorganic carbon system. *Proceedings of the National Academy of Sciences* 104(37): 14580-14585.

Doney, S.C., V.J. Fabry, R.A. Feely, and J.A. Kleypas. 2009. Ocean acidification: The other CO_2 problem. *Annual Review of Marine Science* 1: 169-192.

Dore, J.E., R. Lukas, D.W. Sadler, M.J. Church, and D.M. Karl. 2009. Physical and biogeochemical modulation of ocean acidification in the central North Pacific. *Proceedings of the National Academy of Sciences* 106(30): 12235–12240.

Douvere, F. 2008. The importance of marine spatial planning in advancing ecosystem-based sea use management. *Marine Policy* 32(5): 762-771.

Ducklow, H.W., K. Baker, D.G. Martinson, L.B. Quetin, R.M. Ross, R.C. Smith, S.E. Stammerjohn, M. Vernet, and W. Fraser. 2007. Marine ecosystems: The West Antarctic Peninsula. *Philosophical Transactions of the Royal Society of London B* 362: 67-94.

Ducklow, H., S.C. Doney, and D.K. Steinberg. 2009. Contributions of long-term research and time-series observations to marine ecology and biogeochemistry. *Annual Review of Marine Science* 1: 279-302.

Duda, T. F. 2009. Revisiting experimental methods for studies of acidity-dependent ocean sound absorption. *Journal of the Acoustical Society of America* 125(4): 1971-1981.

Dupont, S. and M.C. Thorndyke. 2009. Impact of CO_2-driven ocean acidification on invertebrates early life-history—What we know, what we need to know and what we can do. *Biogeosciences Discussions* 6: 3109-3131.

Edwards, W. and J.R. Newman. 1982. *Multi-attribute evaluation.* Sage Publications: London.

Egleston, E.S., C.L. Sabine, and F.M.M. Morel. 2010. Revelle revisited: Buffer factors that quantify the response of ocean chemistry to changes in DIC and alkalinity. *Global Biogeochemical Cycles* 24: GB1002.

Engel, A. 2002. Direct relationship between CO_2 uptake and transparent exopolymer particles production in natural phytoplankton. *Journal of Plankton Research* 24(1): 49-53.

Engel A, S. Thoms, U. Riebesell, E. Rochelle-Newall, and I. Zondervan. 2004. Polysaccharide aggregation as a potential sink of marine dissolved organic carbon. *Nature* 428: 929-932.

Enriquez, S., E.R. Mendez, O. Hoegh-Guldberg, and R. Iglesias-Prieto. 2004. The role of multiple scattering by coral skeleton in the amplification of the solar radiation absorbed by coral tissues. *Proceedings of the 10th International Coral Reef Symposium.*

European Science Foundation. 2009. Impacts of ocean acidification. *Science Policy Briefing* 37, August 2009. [Online]. Available: www.esf.org. 12 pp.

Fabry, V.J., C. Langdon, W.M. Balch, A.G. Dickson, R.A. Feely, B. Hales, D.A. Hutchins, J.A. Kleypas, and C.L. Sabine. 2008a. Present and future impacts of ocean acidification on marine ecosystems and biogeochemical cycles. *Report of the Ocean Carbon and Biogeochemistry Scoping Workshop on Ocean Acidification Research* held 9-11 October 2007, La Jolla, CA. 51 pp.

Fabry, V.J., B.A. Seibel, R.A. Feely, and J.C. Orr. 2008b. Impacts of ocean acidification on marine fauna and ecosystem processes. *ICES Journal of Marine Science* 65: 414–432.

Feary, D.A., G.R. Almany, M.I. McCormick, and G.P. Jones. 2007. Habitat choice, recruitment and the response of coral reef fishes to coral degradation. *Oecologia* 153: 727-737.

Feely, R.A. and C.T.A. Chen. 1982. The effect of excess CO_2 on the calculated calcite and aragonite saturation horizons in the northeast Pacific. *Geophysical Research Letters* 9(11): 1294-1297.

Feely, R.A., R.H. Byrne, P.R. Betzer, J.F. Gendron, and J.G. Acker. 1984. Factors influencing the degree of saturation of the surface and intermediate waters of the North Pacific Ocean with respect to aragonite. *Journal of Geophysical Research- Oceans* 89: 10631-10640.

Feely, R.A., R.H. Byrne, J.G. Acker, P.R. Betzer, C-T.A. Chen, J.F. Gendron, and M.F. Lamb. 1988. Winter-summer variations of calcite aragonite saturation in the northeast Pacific. *Marine Chemistry* 25: 227-241.

Feely, R.A., C.L. Sabine, K. Lee, W. Berelson, J. Kleypas, V.J. Fabry, and F.J. Millero. 2004. Impact of Anthropogenic CO_2 on the $CaCO_3$ system in the Oceans. *Science* 305(5682): 362-366.

Feely, R.A., C.L. Sabine, J. Martin Hernandez-Ayon, D. Ianson, and B. Hales. 2008. Evidence for upwelling of corrosive "acidified" water onto the continental shelf. *Science* 320(5882):1490-1492.

Feely, R.A., V.J. Fabry, A. Dickson, J.-P. Gattuso, J. Bijma, U. Riebesell, S. Doney, C. Turley, T. Saino, K. Lee, K. Anthony, and J. Kleypas. 2010. An international observational network for ocean acidification. In *Proceedings of OceanObs'09: Sustained Ocean Observations and Information for Society (Vol. 2)*, Venice, Italy, 21-25 September 2009. Hall, J., D.E. Harrison, and D. Stammer (Eds.). ESA Publication WPP-306.

Fine, M. and D. Tchernov. 2007. Scleractinian coral species survive and recover from decalcification. *Science* 315: 1811.

Fleeger, J.W., K.R. Carman, P.B. Weisenhorn, H. Sofranko, T. Marshall, D. Thistle, and J.P. Barry. 2006. Simulated sequestration of anthropogenic carbon dioxide at a deep-sea site: Effects on nematode abundance and biovolume. *Deep-Sea Research Part I: Oceanographic Research Papers* 53:1135-1147.

Folke, C., S. Carpenter, B. Walker, M. Scheffer, T. Elmqvist, L. Gunderson, and C.S. Holling. 2004. Regime shifts, resilience, and biodiversity in ecosystem management. *Annual Review of Ecology Evolution and Systematics* 35: 557-581.

Food and Agriculture Organization (FAO). 2008. *The state of world fisheries and aquaculture 2008*. Rome 2009.

Frank, K.T., B. Petrie, N.L. Shackell, and J.S. Choi. 2006. Reconciling differences in trophic control in mid-latitude marine ecosystems. *Ecology Letters* 9: 1096-105.

Frank, K.T., B. Petriea, and N.L. Shackell. 2007. The ups and downs of trophic control in continental shelf ecosystems. *Trends in Ecology and Evolution* 22(5): 236-242.

Freedman, A. 2008. Ocean acidification - The sleeper issue. *The Washington Post* July 7, 2008.

Friedlingstein, P., P.M. Cox, R.A. Betts, L. Bopp, W. Von Bloh, V. Brovkin, P. Cadule, S. Doney, M. Eby, I. Fung, G. Bala, J. John, S.D. Jones, F. Joos, T. Kato, M. Kawamiya, W. Knorr, K. Lindsay, H.D. Matthews, T. Raddatz, P. Rayner, C. Reick, E. Roeckner, K-G. Schnitzler, R. Schnur, K. Strassmann, A.J. Weaver, C. Yoshikawa, and N. Zeng. 2006. Climate–carbon cycle feedback analysis: Results from the C^4MIP model intercomparison. *Journal of Climate* 19(14): 3337-3353.

Freiwald, A. 2002. Reef-forming cold-water corals. In *Ocean Margin Systems*. Wefer, G., D. Billett, D. Hebbeln, B.B. Jorgensen, M. Schluter, T. Van Weering (Eds.). Springer, Berlin Heidelberg New York, pp 365–385.

Freiwald, A., J.H. Fosså, A. Grehan, T. Koslow, and J.M. Roberts. 2004. *Cold-water coral reefs*. UNEP-WCMC: Cambridge, UK.

Fu, F-X., M.E. Warner, Y. Zhang, Y. Feng, and D.A. Hutchins. 2007. Effects of increased temperature and CO_2 on photosynthesis, growth, and elemental ratios in marine *synechococcus* and *prochlorococcus* (cyanobacteria). *Journal of Phycology* 43(3): 485-496.

Fu, F-X., M.R. Mulholland, N.S. Garcia, A. Beck, P.W. Bernhardt, M.E. Warner, S.A. Sañudo-Wilhelmy, and D.A. Hutchins. 2008. Interactions between changing pCO_2, N_2 fixation, and Fe limitation in the marine unicellular cyanobacterium *Crocosphaera*. *Limnology and Oceanography* 53(6): 2472-2484.

Furla, P., S. Bénazet-Tambutté, J. Jaubert, and D. Allemand. 1998. Functional polarity of the tentacle of the sea anemone *Anemonia viridis*: Role in inorganic carbon acquisition. *American Journal of Physiology - Regulatory, Integrative and Comparative Physiology* 274: R303–R310.

Gage, J.D. and P.A. Tyler. 1991. *Deep-sea biology: A natural history of organisms at the deep-sea floor*. Cambridge, UK: Cambridge University Press. 504 pp.

Gao, K. and Y. Zheng. 2009. Combined effects of ocean acidification and solar UV radiation on photosynthesis, growth, pigmentation and calcification of the coralline alga *Corallina sessilis* (Rhodophyta). *Global Change Biology* [doi: 10.1111/j.1365-2486.2009.02113.x].

Gardner, W.D., M.J. Richardson, and W.O. Smith, Jr. 2000. Seasonal patterns of water column particulate organic carbon and fluxes in the Ross Sea, Antarctica. *Deep Sea Research Part II: Topical Studies in Oceanography* 47(15-16): 3423-3449.

Gattuso, J-P., M. Frankignoulle, I. Bourge, S. Romaine, and R.W. Buddemeier. 1998. Effect of calcium carbonate saturation of seawater on coral calcification. *Global and Planetary Change* 18(1-2): 37-46.

Gattuso, J-P., L. Hansson, and the EPOCA Consortium. 2009. European Project on Ocean Acidification (EPOCA): Objectives, products, and scientific highlights. *Oceanography* 22(4):190-201.

Gazeau, F., C. Quiblier, J.M. Jansen, J. Gattuso, J.J. Middelberg, and C.H.R. Heip. 2007. Impact of elevated CO_2 on shellfish calcification. *Geophysical Research Letters* L07603. [doi:10.1029/2006GL028554]

Gehlen, M., R. Gangstø, B. Schneider, L. Bopp, O. Aumont, and C. Ethe. 2007. The fate of pelagic $CaCO_3$ production in a high CO_2 ocean: A model study. *Biogeosciences* 4: 505–519.

Gledhill, D.K., R. Wanninkhof, F.J. Millero, and M. Eakin. 2008. Ocean acidification of the Greater Caribbean Region 1996-2006. *Journal of Geophysical Research* 113C10031. [doi:10.1029/2007JC004629]

Glover, D.M., C.L. Chandler, S.C. Doney, K.O. Buesseler, G. Heimerdinger, J.K.B. Bishop, and G.R. Flierl. 2006. The U.S. JGOFS data management experience. *Deep Sea Research Part II: Topical Studies in Oceanography* 53(5-7): 793-802.

Glynn, P.W. 1996. Coral reef bleaching: Facts, hypotheses and implications. *Global Change Biology* 2(6): 495-509.

Goffredi, S.K. and J.J. Childress. 2001. Activity and inhibitor sensitivity of ATPases in the hydrothermal vent tubeworm *Riftia pachyptila*: A comparative approach. *Marine Biology* 138(2): 259-265.

Golomb, D., S. Pennell, D. Ryan, E. Barry, and P. Swett. 2007. Ocean sequestration of carbon dioxide: Modeling the deep ocean release of a dense emulsion of liquid CO_2-in-water stabilized by pulverized limestone particles. *Environmental Science and Technology* 41(13): 4698–4704.

Grabowski, J.H. and C.H. Peterson. 2007. Restoring oyster reefs to recover ecosystem services. In *Ecosystem engineers: Concepts, theory and applications*. Cuddington, K., J.E. Byers, W.G. Wilson, and A. Hastings (Eds.) Elsevier/Academic Press, Netherlands. pp. 281-298.

Graham, N.A.J., S.K. Wilson, S. Jennings, N.V.C. Polunin, J. Robinson, J.P. Bijoux, and T.M. Daw. 2007. Lag effects in the impacts of mass coral bleaching on coral reef fish, fisheries, and ecosystems. *Conservation Biology* 21(5): 1291–1300.

Grassle, J.F. 1986. The ecology of deep-sea hydrothermal vent communities. *Advances in Marine Biology* 23: 301-362.

Gratwicke, B. and M.R. Speight. 2005. The relationship between fish species richness, abundance and habitat complexity in a range of shallow tropical marine habitats. *Journal of Fish Biology* 66: 650-667.

Grebmeier, J.M., J.E. Overland, S.E. Moore, E.V. Farley, E.C. Carmack, L.W. Cooper, K.E. Frey, J.H. Helle, F.A. McLaughlin, and S.L. McNutt. 2006. A major ecosystem shift in the Northern Bering Sea. *Science* 311(5766): 1461-1464.

Green, M.A., M.E. Jones, C.L. Boudreau, R.L. Moore, and B.A. Westman. 2004. Dissolution mortality of juvenile bivalves in coastal marine deposits. *Limnology and Oceanography* 49: 727-734.

Green, M., G. Waldbusser, S. Reilly, K. Emerson, and S. O'Donnel. 2009. Death by dissolution: Sediment saturation state as a mortality factor for juvenile bivalves. *Limnology and Oceanography* 54(4):1037–1047.

Griffiths, C. and W. Wheeler. 2005. Benefit-cost analysis of regulations affecting surface water quality in the United States. In *Cost benefit analysis and water resources management*. Brouwer, R. and D. Pearce (Eds.) Cheltenham, UK: Edward Elgar.

Guinotte, J. and V.J. Fabry. 2009. The threat of ocean acidification to ocean ecosystems. *Journal of Marine Education* 25(1): 2-7.

Guinotte, J.M., J. Orr, S. Cairns, A. Freiwald, L. Morgan, and R. George. 2006. Will human-induced changes in seawater chemistry alter the distribution of deep-sea scleractinian corals?. *Frontiers in Ecology and the Environment* 4(3): 141-146.

Guttal, V. and C. Jayaprakash. 2008. Changing skewness: An early warning signal of regime shifts in ecological systems. *Ecology Letters* 11: 450-460.

Hall-Spencer, J.M., R. Rodolfo-Metalpa, S. Martin, E. Ransome, M. Fine, S.M. Turner, S.J. Rowley, D. Tedesco, and M.C. Buia. 2008. Volcanic carbon dioxide vents show ecosystem effects of ocean acidification. *Nature* 454: 96-99.

Harvey, L.D.D. 2008. Mitigating the atmospheric CO_2 increase and ocean acidification by adding limestone powder to upwelling regions. *Journal of Geophysical Research* 113: C04028.

Havenhand, J.N., F. Buttler, M.C. Thorndyke, and J.E. Williamson. 2008. Near-future levels of ocean acidification reduce fertilization success in a sea urchin. *Current Biology* 18(15): 651-652.

Heinze, C. 2004. Simulating oceanic $CaCO_3$ export production in the greenhouse. *Geophysical Research Letters* 31.

Henderson, J. and L.J. O'Neil. 2003. Economic values associated with construction of oyster reefs by the Corps of Engineers. *EMRRP Technical Notes Collection (ERDC-TN-EMRRP-ER-01)*. Vicksburg, MS: U.S. Army Engineer Research and Development Center. [Online]. Available: http://el.erdc.usace.army.mil/elpubs/pdf/er01.pdf

Hester, K.C., E.T. Peltzer, W.J. Kirkwood, and P.G. Brewer. 2008. Unanticipated consequences of ocean acidification: A noisier ocean at lower pH. *Geophysical Research Letters* 35: L19601.

Heyward, A.J. and A.P. Negri. 1999. Natural inducers for coral larval metamorphosis. *Coral Reefs* 18: 273 279.

Hochachka, P.W. and G.N. Somero. 2002. *Biochemical adaptation: Mechanism and process in physiological evolution*. Oxford University Press: Oxford.

Holland, D., J.N. Sanchirico, R. Johnston and D. Kopklar. 2010. Economic analysis for eco-system based management: Applications to marine and coastal environments. *Resources for the Future Press*. Approx. 250 pages. [Online]. Available: http://www.earthscan.co.uk/?tabid=102257

Hönisch, B. 2005. Surface ocean pH response to variations in pCO_2 through two full glacial cycles. *Earth and Planetary Science Letters* 236(1-2): 305-314.

Hönisch, B., N.G. Hemming, D. Archer, M. Siddall, and J.F. McManus. 2009. Atmospheric carbon dioxide concentration across the mid-pleistocene transition. *Science*. 324(5934): 1551-1554.

Honjo, S., R. Francois, S. Manganini, J. Dymond, and R. Collier. 2000. Particle fluxes to the interior of the Southern Ocean in the Western Pacific sector along 170°W. *Deep Sea Research Part II: Topical Studies in Oceanography* 47(15-16): 3521-3548.

Hopkins F.E., S.M. Turner, P.D. Nightingale, M. Steinke, D. Bakker, and P.S. Liss. 2010. Ocean acidification and marine trace gas emissions. *Proceedings of the National Academy of Sciences* 107: 760-765.

Hossain, M.M.M. and S. Ohde. 2006. Calcification of cultured *Porites* and *Fungia* under different aragonite saturation states of seawater. *Proceedings of the 10th International Coral Reef Symposium. Japan Coral Reef Society* 597–606.

House, K.Z., D.P. Schrag, C.F. Harvey, and K.S. Lackner. 2006. Permanent carbon dioxide storage in deep-sea sediments. *Proceedings of the National Academy of Sciences* 103(33): 12291-12295.

House, K.Z., C.H. House, D.P. Schrag, and M.J. Aziz. 2007. Electrochemical acceleration of chemical weathering as an energetically feasible approach to mitigating anthropogenic climate change. *Environmental Science & Technology* 41: 8464-8470.

Hutchins, D.A., F.X. Fu, Y. Zhang, M.E. Warner, Y. Feng, K. Portune, P.W. Bernhardt, and M.R. Mulholland. 2007. CO_2 control of Trichodesmium N_2 fixation, photosynthesis, growth rates, and elemental ratios: Implications for past, present, and future ocean biogeochemistry. *Limnology and Oceanography* 52(4): 1293-1304.

Hutchins, D.A., M.R. Mulholland, and F. Fu. 2009. Nutrient cycles and marine microbes in a CO_2-enriched ocean. *Oceanography* 22(4): 128-145.

Iglesias-Rodriguez, D.M., P.R. Halloran, R.E.M. Rickaby, I.R. Hall, E. Colmenero-Hidalgo, J.R. Gittins, D.R.H. Green, T. Tyrrell, S.J. Gibbs, P. von Dassow, E. Rehm, E.V. Armbrust, and K.P. Boessenkool. 2008. Phytoplankton calcification in a high-CO_2 world. *Science* 320: 336-340.

Intergovernmental Panel on Climate Change (IPCC). 2000. *Special Report on Emissions Scenarios*. Cambridge University Press: Cambridge, UK and New York, NY, USA.

Integrated Ocean Observing System. 2009. U.S. IOOS®: Our eyes on our oceans, coasts, and Great Lakes. [Online]. Available: http://ioos.gov/ [Accessed on December 2, 2009]

Johnson, J.E. and P.A. Marshall (Eds). 2007. *Climate change and the Great Barrier Reef*. Great Barrier Reef Marine Park Authority and Australian Greenhouse Office. Townsville, Australia.

Joint, I., D.M. Karl, S.C. Doney, E.V. Armbrust, W. Balch, M. Berman, C. Bowler, M. Church, A. Dickson, J. Heidelberg, D. Iglesias-Rodriguez, D. Kirchman, Z. Kolber, R. Letelier, C. Lupp, S. Maberly, S. Park, J. Raven, D.J. Repeta, U. Riebesell, G. Steward, P. Tortell, R.E. Zeebe, and J.P. Zehr. 2009. *Consequences of high CO_2 and ocean acidification for microbes in the global ocean*. Report of expert meeting at the University of Hawaii, 24-26 February 2009 organized by Plymouth Marine Laboratory and Center for Microbial Oceanography Research and Education. 23pp.

Jokiel, P.L. and S.L. Coles. 1977. Effects of temperature on mortality and growth of Hawaiian reef corals. *Marine Biology* 43: 201-208.

Jokiel, P.L., K.S. Rodgers, I.B. Kuffner, A.J. Andersson, E.F. Cox, and F.T. Mackenzie. 2008. Ocean acidification and calcifying reef organisms: A mesocosm investigation. *Coral Reefs* 27: 473-483.

Keeney, R. 1992. *Value-focused thinking: A path to creative decisionmaking.* Harvard University Press: Cambridge, MA.

Keller, B.D., S. Airamé, B. Causey, A. Friedlander, D.F. Gleason, R. Grober-Dunsmore, J. Johnson, E. McLeod, S.L. Miller, R.S. Steneck, and C. Woodley. 2008. Marine protected areas. In: *Preliminary review of adaptation options for climate-sensitive ecosystems and resources.* A Report by the U.S. Climate Change Science Program and the Subcommittee on Global Change Research [Julius, S.H., J.M. West (Eds.), J.S. Baron, B. Griffith, L.A. Joyce, P. Kareiva, B.D. Keller, M.A. Palmer, C.H. Peterson, and J.M. Scott (Authors)]. U.S. Environmental Protection Agency. [Online]. Available: http://downloads.climate-science.gov/sap/sap4-4/sap4-4-final-report-Ch8-mpa.pdf [Accessed October 4, 2009]

Kelly, P.M. and W.N. Adger. 2000. Theory and practice in assessing vulnerability to climate change and facilitating adaptation. *Climate Change* 47: 325-252.

Kennett, J.P. and L.D. Stott. 1991. Abrupt deep sea warming, paleoceanographic changes and benthic extinctions at the end of the Paleocene. *Nature* 353: 319-322.

Key, R.M., A. Kozyr, C.L. Sabine, K. Lee, R. Wanninkhof, J.L. Bullister, R.A. Feely, F.J. Millero, C. Mordy, and T-H. Peng. 2004. A global ocean carbon climatology: Results from Global Data Analysis Project (GLODAP). *Global Biogeochemical Cycles* 18: GB4031.

Khatiwala, S., F. Primeau, and T. Hall. 2009. Reconstruction of the history of anthropogenic CO_2 concentrations in the ocean. *Nature* 462: 346-349.

Kiessling, W., C. Simpson, and M. Foote. 2010. Reefs as cradles of evolution and sources of biodiversity in the Phanerozoic. *Science* 327(5962): 196-198.

Kim, T.Y., S. Kwak, and S. Yoo. 1998. Applying multi-attribute utility theory to decision making in environmental planning: A case study of the electric utility in Korea. *Journal of Environmental Planning and Management* 41(5): 597-609.

Kim, J-M., K. Lee, K. Shin, J-H. Kang, H-W. Lee, M. Kim, P-G. Jang, and M-C. Jang. 2006. The effect of seawater CO_2 concentration on growth of a natural phytoplankton assemblage in a controlled mesocosm experiment. *Limnology and Oceanography* 51(4): 1629-1636.

Kim, J-M., K. Shin, K. Lee, and B-K. Park. 2008. *In situ* ecosystem-based carbon dioxide perturbation experiments: Design and performance evaluation of a mesocosm facility. *Limnology and Oceanography: Methods* 6: 208–217.

Kitchingman, A. and S. Lai. 2004. Inferences of potential seamount locations from mid-resolution bathymetric data. In Morato and Pauly (Eds.). *Seamounts: Biodiversity and fisheries. Fisheries Centre Research Reports* 12(5): 7-12.

Klaas, C. and D.E. Archer. 2002. Association of sinking organic matter with various types of mineral ballast in the deep sea: Implications for the rain ratio. *Global Biogeochemistry Cycles* 16(4): 1116.

Kleinen, T., H. Held, and G. Petschel-Held. 2003. The potential role of spectral properties in detecting thresholds in the earth system: Application to the thermohaline circulation. *Ocean Dynamics* 53: 53-63.

Kleypas, J.A., R.A. Feely, V.J. Fabry, C. Langdon, C.L. Sabine, and L.L. Robbins. 2006. *Impacts of ocean acidification on coral reefs and other marine calcifiers: A guide for future research.* Report of a workshop held 18-20 April 2005, St. Petersburg, FL. 88 pp.

Kleypas, J.A. and C. Langdon. 2006. Coral reefs and changing seawater carbonate chemistry. *Coral Reefs and Climate Change: Science and Management Coastal and Estuarine Studies,* American Geophysical Union 61: 73-100.

Kling, D. and J.N. Sanchirico. 2009. An adaptation portfolio for the United States coastal and marine environments. *Resources for the Future Report.* [Online]. Available: www.rff.org

Knoll, A.H., R.K. Bambach, D.E. Canfield, and J.P. Grotzinger. 1996. Comparative Earth history and late Permian mass extinction. *Science* 273: 452-457.

Kranz, S.A., D. Sultemeyer, K.U. Richter, and B. Rost. 2009. Carbon acquisition by Trichodesmium: The effect of pCO_2 and diurnal changes. *Limnology and Oceanography* 54(2): 548-559.

Kuffner, I.B., A.J. Andersson, P.L. Jokiel, K.S. Rodgers, and F.T. Mackenzie. 2008. Decreased abundance of crustose coralline algae due to ocean acidification. *Nature Geoscience* 1: 77-140.

Kurihara, H. 2008. Effects of CO_2-driven ocean acidification on the early developmental stages of invertebrates. *Marine Ecological Progress Series* 373: 275–284.

Kurihara, H. and Y. Shirayama. 2004. Effects of increased atmospheric CO_2 on sea urchin early development. *Marine Ecology Progress Series* 274: 161–169.

Kurihara, H., S. Kato, and A. Ishimatsu. 2007. Effects of increased seawater pCO_2 on early development of the oyster *Crassostrea gigas*. *Aquatic Biology* 1: 91-98.

Kurihara, H., T. Asai, S. Kato, and A. Ishimatsu. 2008a. Effects of elevated pCO_2 on early development in the mussel *Mytilus galloprovincialis*. *Aquatic Biology* 4: 225–233.

Kurihara, H., M. Matsui, H. Furukawa, M. Hayashi, and A. Ishimatsu. 2008b. Long-term effects of predicted future seawater CO_2 conditions on the survival and growth of the marine shrimp *Palaemon pacificus*. *Journal of Experimental Marine Biology and Ecology* 367: 41-46.

Langdon, C. and M.J. Atkinson. 2005. Effect of elevated pCO_2 on photosynthesis and calci-fication of corals and interactions with seasonal change in temperature/irradiance and nutrient enrichment. *Journal of Geophysical Research: Oceans* 110: C09S07. [doi:10.1029/2004JC002576]

Langdon, C., T. Takahashi, C. Sweeney, D. Chipman, J. Goddard, F. Marubini, H. Aceves, H. Barnett, and M.J. Atkinson. 2000. Effect of calcium carbonate saturation state on the cal-cification rate of an experimental coral reef. *Global Biogeochemical Cycles* 14: 639-654.

Langdon, C., W.S. Broecker, D.E. Hammond, E. Glenn, K. Fitzsimmons, S.G. Nelson, T.H. Peng, I. Hajdas, and G. Bonani. 2003. Effect of elevated CO_2 on the community metabo-lism of an experimental coral reef. *Global Biogeochemical Cycles* 17: 1011. [doi:10.1029/2002GB001941]

Langer G., M. Geisen, K. Baumann, J. Kläs, U. Riebesell, S. Thoms, and J.R. Young. 2006. Species-specific responses of calcifying algae to changing seawater carbonate chemistry. *Geochemistry Geophysics Geosystems* 7: Q09006. [doi:10.1029/2005GC001227]

Langer, G., G. Nehrke, I. Probert, J. Ly, and P. Ziveri. 2009. Strain-specific responses of *Emiliania huxleyi* to changing seawater carbonate chemistry. *Biogeosciences Discussions* 6: 4361–4383.

Leclercq N., J.-P. Gattuso, and J. Jaubert. 2000. CO_2 partial pressure controls the calcification rate of a coral community. *Global Change Biology* 6: 329-334.

Leclercq, N., J-P. Gattuso, and J. Jaubert. 2002. Primary production, respiration, and calcifi-cation of a coral reef mesocosm under increased CO_2 partial pressure. *Limnology and Oceanography* 47(2): 558-564.

Lee, K., L.T. Tong, F.J. Millero, C.L. Sabine, A.G. Dickson, C. Goyet, G-H. Park, R. Wanninkhof, R.A. Feely, and R.M. Key. 2006. Global relationships of total alkalinity with salinity and temperature in surface waters of the world's oceans. *Geophysical Research Letters* 33: L19605. [doi:10.1029/2005GL027207]

Levitan, O., G. Rosenberg, I. Setlik, E. Setlikova, J. Grigel, J. Klepetar, O. Prasil, and I. Berman-Frank. 2007. Elevated CO_2 enhances nitrogen fixation and growth in the marine cyanobacterium Trichodesmium. *Global Change Biology* 13: 531-538.

Lewis, E. and D.W.R. Wallace. 1998. *Program developed for CO_2 system calculations*. ORNL/CDIAC-105. Carbon Dioxide Information Analysis Center, Oak Ridge National Labora-tory, U.S. Department of Energy: Oak Ridge, Tennessee.

Liu, Y. 2009. Instability of seawater pH in the South China Sea during the mid-late Holocene: Evidence from boron isotopic composition of corals. *Geochimica Et Cosmochimica Acta* 73(5): 1264-1272.

Lourens, L.J., A. Sluijs, D. Kroon, J.C. Zachos, E. Thomas, U. Rohl, J. Bowles, and I. Raffi. 2005. Astronomical pacing of late Palaeocene to early Eocene global warming events. *Nature* 435: 1083-1087.

Lueker, T.J., A.G. Dickson, and C.D. Keeling. 2000. Ocean pCO_2 calculated from dissolved inorganic carbon, alkalinity, and equations for K_1 and K_2: Validation based on laboratory measurements of CO_2 in gas and seawater at equilibrium. *Marine Chemistry* 70(1-3): 105-119.

Maier, C., J. Hegeman, M.G. Weinbauer, and J.P. Gattuso. 2009. Calcification of the cold-water coral *Lophelia pertusa* under ambient and reduced pH. *Biogeosciences Discussions* 6: 1875-1901.

Manzello, D.P., J.A. Kleypas, D.A. Budd, C.M. Eakin, P.W. Glynn, and C. Langdon. 2008. Poorly cemented coral reefs of the eastern tropical Pacific: Possible insights into reef development in a high-CO_2 world. *Proceedings of the National Academy of Sciences* 105(30): 10450-10455.

Marshall, A.T. and P.L. Clode. 2002. Effect of increased calcium concentration in sea water on calcification and photosynthesis in the scleractinian coral *Galaxea fascicularis*. *Journal of Experimental Biology* 205: 2107–2113.

Martin, S. and J-P. Gattuso. 2009. Response of Mediterranean coralline algae to ocean acidification and elevated temperature. *Global Change Biology* 15(8): 2089-2100. [doi: 10.1111/j.1365-2486.2009.01874.x]

Martin, S., R. Rodolfo-Metalpa, E. Ransome, S. Rowley, M.C. Buia, J.P. Gattuso, and J. Hall-Spencer. 2008. Effects of naturally acidified seawater on seagrass calcareous epibionts. *Biology Letters* 4: 689-692.

Martz, T.R., K.S. Johnson, H. Jannasch, L. Coletti, J. Barry, and C. Lovera. 2008. *ISFET sensor evaluation and modification for seawater pH measurement*. American Geophysical Union Fall Meeting. abstract #OS33E-02.

Marubini, F. and M.J. Atkinson. 1999. Effects of lowered pH and elevated nitrate on coral calcification. *Marine Ecology Progress Series* 188: 117-121.

Marubini, F., H. Barnett, C. Langdon, and M.J. Atkinson. 2001. Dependence of calcification on light and carbonate ion concentration for the hermatypic coral *Porites compressa*. *Marine Ecology Progress Series* 220: 153-162.

Marubini, F., C. Ferrier-Pages, and J-P. Cuif. 2003. Suppression of skeletal growth in scleractinian corals by decreasing ambient carbonate-ion concentration: A cross-family comparison. *Proceedings of the Royal Society B: Biological Sciences* 270: 179-184.

Marubini, F., C. Ferrier-Pages, P. Furla, and D. Allemand. 2008. Coral calcification responds to seawater acidification: A working hypothesis towards a physiological mechanism. *Coral Reefs* 27: 491-499.

McDonald, M.R., J.B. McClintock, C.D. Amsler, D. Rittschof, R.A. Angus, B. Orihuela, and K. Lutostanski. 2009. Effects of ocean acidification over the life history of the barnacle *Amphibalanus amphitrite*. *Marine Ecology Progress Series* 385: 179-187.

McLeod, K. and H. Leslie. (Eds.). 2009. *Ecosystem-Based Management for the Oceans*. Island Press: Washington, D.C. 392 pp.

McGinn, P.J. and F.M.M. Morel. 2008. Expression and inhibition of the carboxylating and decarboxylating enzymes in the photosynthetic C_4 pathway of marine diatoms. *Plant Physiology* 146: 1-10.

Melzner, F., M.A. Gutowska, M. Langenbuch, S. Dupont, M. Lucassen, M.C. Thorndyke, M. Bleich, and H-O. Pörtner. 2009. Physiological basis for high CO_2 tolerance in marine ectothermic animals: Pre-adaptation through lifestyle and ontogeny?. *Biogeosciences Discussions* 6(3): 4693-4738.

Meseck, S.L., B.C. Smith, G.H. Wikfors, J.H. Alix, and D. Kapareiko. 2007. Nutrient interactions between phytoplankton and bacterioplankton under different carbon dioxide regimes. *Journal of Applied Phycology* 19(3): 229-237.

Meyer, D.L., E.C. Townsend, and G.W. Thayer. 1997. Stabilization and erosion control value of oyster clutch for intertidal marsh. *Restoration Ecology* 5: 93-99.

Miles, E.L. 2009. On the increasing vulnerability of the world ocean to multiple stresses. *Annual Review of Environment and Resources* 34: 17-41.

Miller, A.W., A.C. Reynolds, C. Sobrino, and G.F. Riedel. 2009. Shellfish face uncertain future in high CO_2 world: Influence of acidification on oyster larvae calcification and growth in estuaries. *PLoS ONE* 4(5): e5661. [doi:10.1371/journal.pone.0005661]

Millero, F.J. 2006. *Chemical oceanography*. 3rd Edition. CRC Press: Boca Raton, FL. 496 p.

Millero, F.J., R. Woosley, B. DiTrolio, and J. Waters. 2009. The effect of ocean acidification on the speciation of metals in seawater. *Oceanography* 22(4): 72-85.

Morel, F.M.M. and J.G. Hering. 1993. *Principles and applications of aquatic chemistry.* John Wiley: New York, NY.

Morel, F.M.M., A.J. Milligan, and M.A. Saito. 2003. Marine bioinorganic chemistry: The role of trace metals in the oceanic cycles of major nutrients. In *Treatise on Geochemistry.* K.K. Turekian, H.D. Holland, (Eds.). Elsevier Science Ltd.: Cambridge, U.K. 6: 113-143.

Morse, J.W., A.J. Andersson, and F.T. Mackenzie. 2006. Initial responses of carbonate-rich shelf sediments to rising atmospheric pCO_2 and "ocean acidification": Role of high Mg-calcites. *Geochimica et Cosmochimica Acta* 70: 5814-5830.

Morse, J.W. and F.T. Mackenzie. 1990. *Geochemistry of sedimentary carbonates.* Elsevier: Amsterdam. 707 pp.

Mortensen, P.B. and L. Buhl-Mortensen. 2005. Morphology and growth of the deep-water gorgonians *Primnoa resedaeformis* and *Paragorgia arborea. Marine Biology* 147: 775-788.

Moy, A.D., W.R. Howard, S.G. Bray, and T.W. Trull. 2009. Reduced calcification in modern Southern Ocean planktonic foraminifera. *Nature Geoscience* 2: 276-280. [doi:10.1029/NGEO460]

Munday, P.L., D.L. Dixson, J.M. Donelson, G.P. Jones, M.S. Pratchett, G.V. Devitsina, and K.B. Doving. 2009. Ocean acidification impairs olfactory discrimination and homing ability of a marine fish. *Proceedings of the National Academy of Sciences* 106: 1848-1852.

National Oceanic and Atmospheric Administration. 2008. *Fisheries of the United States 2007.* Current Fishery Statistics No. 2007. National Marine Fisheries Service, Office of Science and Technology. Silver Spring, MD.

National Oceanic and Atmospheric Administration. 2009a. *Marine invertebrates and plants.* Office of Protected Resources, NOAA Fisheries. [Online]. Available: http://www.nmfs.noaa.gov/pr/species/invertebrates/#corals [Accessed October 4, 2009]

National Oceanic and Atmospheric Administration. 2009b. *A vision for climate services in NOAA.* [Online]. Available: http://www.climate.noaa.gov/pdf/GandPdocumentOct21.pdf [Accessed on December 29, 2009]

National Research Council. 1996. *Understanding Risk: Informing Decisions in a Democratic Society.* National Academy Press, Washington, D.C.

National Research Council. 1999. *Global Ocean Science: Toward an Integrated Approach.* National Academy Press, Washington, D.C.

National Research Council. 2005a. *Decision Making for the Environment: Social and Behavioral Science Research Priorities.* National Academies Press, Washington, D.C.

National Research Council. 2005b. *Thinking Strategically: The Appropriate Use of Metrics for the Climate Change Science Program.* National Academies Press, Washington, D.C.

National Research Council. 2007a. *Analysis of Global Change Assessments: Lessons Learned.* National Academies Press, Washington, D.C.

National Research Council. 2007b. *Evaluating Progress of the U.S. Climate Change Science Program: Methods and Preliminary Results.* National Academies Press, Washington, D.C.

National Research Council. 2007c. *Research and Networks for Decision Support in the NOAA Sectoral Applications Research Program*. National Academies Press, Washington, D.C.

National Research Council. 2008. *Public Participation in Environmental Assessment and Decision Making*. National Academies Press, Washington, D.C.

National Research Council. 2009a. *Informing Decisions in a Changing Climate*. National Academies Press, Washington, D.C.

National Research Council. 2009b. *Restructuring Federal Climate Research to Meet the Challenges of Climate Change*. National Academies Press, Washington, D.C.

National Science Foundation. 2009. *National Science Foundation Award Search*. [Online]. Available: http://www.nsf.gov/awardsearch/ [Accessed on December 29, 2009]

Negri, A.P., N.S. Webster, R.T. Hill, and A.J. Heyward. 2001. Metamorphosis of broadcast spawning corals in response to bacteria isolated from crustose algae. *Marine Ecology Progress Series* 223: 121-131.

Newell, R.I.E. 1988. Ecological changes in Chesapeake Bay: Are they the result of over-harvesting the American Oyster *Crassostrea virginica*? In *Understanding the estuary: Advances in Chesapeake Bay research*. (Eds.) Lynch, M. P. and E.C. Krome. Publication 129. Chesapeake Research Consortium: Baltimore, MD.

Newell, R.I.E. and E.W. Koch. 2004. Modeling seagrass density and distribution in response to change in tubidity stemming from bivalve filtration and seagrass sediment stabilization. *Estuaries* 27: 793-806.

Nordhaus, W.D. 2007. A review of the stern review on the economics of climate change. *Journal of Economic Literature* 45(3): 686-702.

Norström, A.V., M. Nyström, J. Lokrantz, and C. Folke. 2009. Alternative states on coral reefs: Beyond coral-macroalgal phase shifts. *Marine Ecology Progress Series* 376: 295-306.

Ocean Carbon and Biogeochemistry Program. 2009a. *Ocean acidification—recommended strategy for a U.S. national research program*. Ocean Carbon and Biogeochemistry Program. 14 pp. [Online]. Available: http://us-ocb.org/.

Ocean Carbon and Biogeochemistry Program. 2009b. *Ocean carbon and biogeochemistry program response to EPA notice of data availability: Ocean acidification and marine pH water quality criteria*. Public Submission EPA-HQ-OW-2009-0224-0163, 33pp. [Online]. Available: http://www.regulations.gov/fdmspublic/component/main?main=DocumentDetail&o=09000064809d0189, http://us-ocb.org/.

Ohde, S. and M.M.M. Hossain. 2004. Effect of $CaCO_3$ (aragonite) saturation state of seawater on calcification of *Porites* coral. *Geochemical Journal* 38: 613-621.

Olafsson J., S.R. Olafsdottir, A. Benoit-Cattin, M. Danielsen, T.S. Arnarson, and T. Takahashi. 2009. Rate of Iceland Sea acidification from time series measurements. *Biogeosciences* 6: 2661-2668.

Orr, J.C., V.J. Fabry, O. Aumont, L. Bopp, S.C. Doney, R.A. Feely, A. Gnanadesikan, N. Gruber, A. Ishida, F. Joos, R.M. Key, K. Lindsay, E. Maier-Reimer, R. Matear, P. Monfray, A. Mouchet, R.G. Najjar, G-K. Plattner, K.B. Rodgers, C.L. Sabine, J.L. Sarmiento, R. Schlitzer, R.D. Slater, I.J. Totterdell, M-F. Weirig, Y. Yamanaka, and A. Yool. 2005. Anthropogenic ocean acidification over the twenty-first century and its impact on calcifying organisms. *Nature* 437: 681-686.

Orr, J.C., K. Caldeira, V. Fabry, J-P. Gattuso, P. Haugan, P. Lehodey, S. Pantoja, H-O. Pörtner, U. Riebesell, T. Trull, E. Urban, M. Hood, and W. Broadgate. 2009. Research priorities for understanding ocean acidification. *Oceanography* 22: 182-189.

Oschlies, A., K.G. Schulz, U. Riebesell, and A. Schmittner. 2008. Simulated 21st century's increase in oceanic suboxia by CO_2-enhanced biotic carbon export. *Global Biogeochemical Cycles* 22: GB4008. [doi:10.1029/2007GB003147]

Overland, J., S. Rodionov, S. Minobe, and N. Bond. 2008. North Pacific regime shifts: Definitions, issues and recent transitions. *Progress in Oceanography* 77: 92-102.

Pacala, S. and R. Socolow. 2004. Stabilization wedges: Solving the climate problem for the next 50 years with current technologies. *Science* 305(5686): 968-972.

Padan, E., E. Bibi, M. Ito, and T.A. Krulwich. 2005. Alkaline pH homeostasis in bacteria: New insights. *Biochimica et Biophysica Acta - Biomembranes* 1717(2): 67-88.

Paddack, M., J. Reynolds, C. Aguilar, R. Appeldoorn, J. Beets, E. Burkett, P. Chittaro, K. Clarke, R. Esteves, and A. Fonseca. 2009. Recent region-wide declines in Caribbean reef fish abundance. *Current Biology* 19(7): 590-595.

Pagani, M., K. Caldeira, D. Archer, and J.C. Zachos. 2006. An ancient carbon mystery. *Science* 314 (5805): 1556-1557.

Pala, C. 2009. The Pacific Ocean's acidification laboratory. *Environmental Science and Technology* 43: 6451-6452.

Palacios, S.L., and R.C. Zimmerman. 2007. Response of eelgrass *Zostera marina* to CO_2 enrichment: Possible impacts of climate change and potential for remediation of coastal habitats. *Marine Ecology Progress Series* 344: 1-13.

Palenik, B. and F.M.M. Morel. 1991a. Comparison of cell-surface L-amino acid oxidases from several marine phytoplankton. *Marine Ecology Progress Series* 59: 195-201.

Palenik, B. and F.M.M. Morel. 1991b. Amine oxidases of marine phytoplankton. *Applied and Environmental Microbiology* 57: 2440-2443.

Palenik, B. and F.M.M. Morel. 1990. Amino acid utilization by marine phytoplankton: A novel mechanism. *Limnology and Oceanography* 35: 260-269.

Palumbi, S.R., P.A. Sandifer, J.D. Allan, M.W. Beck, D.G. Fautin, M.J. Fogarty, B.S. Halpern, L.S. Incze, J. Leong, E. Norse, J.J. Stachowicz, and D.H. Wall. 2009. Managing for ocean biodiversity to sustain marine ecosystem services. *Frontiers in Ecology and the Environment* 7(4): 204-211.

Pane, E.F. and J.P. Barry. 2007. Extracellular acid–base regulation during short-term hypercapnia is effective in a shallow-water crab, but ineffective in a deep-sea crab. *Marine Ecology Progress Series* 334: 1-9.

Parker, L.M., P.M. Ross, and W.A. O'Connor. 2009. The effect of ocean acidification and temperature on the fertilization and embryonic development of the Sydney rock oyster *Saccostrea glomerata* (Gould 1850). *Global Change Biology* 15(9): 2123-2136. [doi: 10.1111/j.1365-2486.2009.01895.x]

Pearse, V., J. Pearse, M. Buchsbaum, and R. Buchsbaum. 1987. *Living Invertebrates*. Blackwell Scientific Pub.: Boston, MA.

Petit, J.R., J. Jouzel, D. Raynaud, N.I. Barkov, J.M. Barnola, I. Basile, M. Bender, J. Chappellaz, J. Davis, G. Delaygue, M. Delmotte, V.M. Kotlyakov, M. Legrand, V. Lipenkov, C. Lorius, L. Pépin, C. Ritz, E. Saltzman, and M. Stievenard. 1999. Climate and atmospheric history of the past 420,000 years from the Vostok Ice Core, Antarctica. *Nature* 399: 429-436.

Pew Center for Global Climate Change. 2009. *The science and consequences of ocean acidification.* Science Brief 3. 8pp. [Online]. Available: http://www.pewclimate.org/docUploads/ocean-acidification-Aug2009.pdf

Politi, Y., T. Arad, E. Klein, S. Weiner, and L. Addadi. 2004. Sea urchin spine calcite forms via a transient amorphous calcium carbonate phase. *Science* 306(5699): 1161-1164.

Pörtner, H.O. and A.P. Farell. 2008. Physiology and climate change. *Science* 322(5902): 690-692. [doi:10.1126/science.1163156]

Pörtner, H.O., C. Bock, and A. Reipschläger. 2000. Modulation of the cost of pHi regulation during metabolic depression: A ^{31}P-NMR study in invertebrate (*Sipunculus nudus*) isolated muscle. *The Journal of Experimental Biology* 203: 2417-2428.

Pörtner, H.O., M. Langenbuch, and A. Reipschläger. 2004. Biological impact of elevated ocean CO_2 concentrations: Lessons from animal physiology and Earth history. *Journal of Oceanography* 60: 705-718.

Pratchett, M.S., P.L. Munday, S.K. Wilson, N.A.J. Graham, J.E. Cinner, D.R. Bellwood, G.P. Jones, N.V.C. Polunin, and T.R. McClanahan. 2008. Effects of climate-induced coral bleaching on coral-reef fishes- Ecological and economic consequences. *Oceanography and Marine Biology: An Annual Review* 46: 251-296.

Ramsey, F.P. 1928. A mathematical theory of saving. *Economic Journal* 38(152): 543–59.

Rau, G.H. 2008. Electrochemical splitting of calcium carbonate to increase solution alkalinity: Implications for mitigation of carbon dioxide and ocean acidity. *Environmental Science and Technology* 42(23):8935-8940.

Rau, G.H. and K. Caldeira. 1999. Enhanced carbonate dissolution: A means of sequestering waste CO_2 as ocean bicarbonate. *Energy Conversion And Management* 40: 1803-1813.

Raven, J., K. Caldeira, H. Elderfield, O. Hoegh-Guldberg, P.S. Liss, U. Reisbell, J. Shepard, C. Turley, and A.J. Watson. 2005. *Ocean acidification due to increasing atmospheric carbon dioxide*. Policy Document. The Royal Society, London.

Reinfelder, J.R., A.M.L. Kraepiel, and F.M.M. Morel. 2000. Unicellular C_4 photosynthesis in a marine diatom. *Nature* 407: 996-999.

Reinfelder, J.R., A.J. Milligan, and F.M.M. Morel. 2004. The role of the C_4 pathway in carbon accumulation and fixation in a marine diatom. *Plant Physiology* 135: 2106-2111.

Renegar, D.A. and B.M. Riegl. 2005. Effect of nutrient enrichment and elevated CO_2 partial pressure on growth rate of Atlantic scleractinian coral *Acropora cervicornis*. *Marine Ecology Progress Series* 293: 69-76.

Reynaud, S., N. Leclercq, S. Romaine-Lioud, C. Ferrier-Pages, J. Jaubert, and J.P. Gattuso. 2003. Interacting effects of CO_2 partial pressure and temperature on photosynthesis and calcification in a scleractinian coral. *Global Change Biology* 9: 1660-1668.

Ridgwell, A. and R.E. Zeebe. 2005. The role of the global carbonate cycle in the regulation and evolution of the Earth system. *Earth and Planetary Science Letters* 234(3-4): 299-315.

Ridgwell, A., D.N. Schmidt, C. Turley, C. Brownlee, M.T. Maldonado, P. Tortell, and J.R. Young. 2009. From laboratory manipulations to earth system models: Predicting pelagic calcification and its consequences. *Biogeosciences Discussions* 6: 3455–3480.

Riebesell, U. 2008. Climate change: Acid test for marine biodiversity. *Nature* 454: 46-47.

Riebesell, U., I. Zondervan, B. Rost, P.D. Tortell, R.E. Zeebe, and F.M.M. Morel. 2000. Reduced calcification in marine plankton in response to increased atmospheric CO_2. *Nature* 407: 634-637.

Riebesell, U., K.G. Schulz, R.G.J. Bellerby, M. Botros, P. Fritsche, M. Meyerhöfer, C. Neill, G. Nondal, A. Oschlies, J. Wohlers, and E. Zöllner. 2007. Enhanced biological carbon consumption in a high CO_2 ocean. *Nature* 450: 545-548.

Riebesell, U., R.G.J. Bellerby, A. Engel, V.J. Fabry, D.A. Hutchins, T.B.H. Reusch, K.G. Schulz, and F.M.M. Morel. 2008. Comment on "Phytoplankton calcification in a high-CO_2 world". *Science* 322: 1466b.

Riebesell, U., A. Körtzinger, and A. Oschlies. 2009. Sensitivities of marine carbon fluxes to ocean change. *Proceedings of the National Academy of Sciences* 106: 20602-20609.

Riebesell, U., V.J. Fabry, L. Hansson, and J-P. Gattuso (Eds). 2010. *Guide to best practices in ocean acidification research and data reporting*. Office for Official Publications of the European Communities. Luxembourg: 264 pp.

Ries, J.B., A.L. Cohen, and D.C. McCorkle. 2009. Marine calcifiers exhibit mixed responses to CO_2-induced ocean acidification. *Geology* 37(12): 1131-1134.

Rietkerk, M., S.C. Dekker, P.C. de Ruiter, and J. van de Koppel. 2004. Self-organized patchiness and catastrophic shifts in ecosystems. *Science* 305: 1926-1929.

Roark, E.B., T.P. Guilderson, R.B. Dunbar, S.J. Fallon, and D.A. Mucciarone. 2009. Extreme longevity in proteinaceous deep-sea corals. *Proceedings of the National Academy of Sciences* 106(13): 5204-5208.

Roberts, J.M., A.J. Wheeler, and A. Freiwald. 2006. Reefs of the deep: The biology and geology of cold-water coral ecosystems. *Science* 312: 543-547.

Rosenzweig, C., D. Karoly, M. Vicarelli, P. Neofotis, Q. Wu, G. Casassa, A. Menzel, T.L. Root, N. Estrella, B. Seguin, P. Tryjanowski, C. Liu, S. Rawlins, and A. Imeson. 2008. Attributing physical and biological impacts to anthropogenic climate change. *Nature* 453(7193): 353-358.

Rost, B. and U. Riebesell. 2004. Coccolithophores and the biological pump: Responses to environmental changes. In: *Coccolithophores – From Molecular Processes to Global Impact.* H.R. Thierstein, J.R. Young (Eds.). Springer: New York, NY. 76-99.

Rost, B., I. Zondervan, and D. Wolf-Gladrow. 2008. Sensitivity of phytoplankton to future changes in ocean carbonate chemistry: Current knowledge, contradictions and research directions. *Marine Ecology Progress Series* 373: 227-237.

Russell, B.D., J.I. Thompson, L.J. Falkenberg, and S.D. Connell. 2009. Synergistic effects of climate change and local stressors: CO_2 and nutrient-driven change in subtidal rocky habitats. *Global Change Biology* 15: 2153-2162.

Sabine, C.L., R.A. Feely, N. Gruber, R.M. Key, K. Lee, J.L. Bullister, R. Wanninkhof, C.S. Wong, D.W.R. Wallace, B. Tilbrook, F.J. Millero, T-H. Peng, A. Kozyr, T. Ono, and A.F. Rios. 2004. The oceanic sink for anthropogenic CO_2. *Science* 305(5682): 367-371.

Salisbury, J., M. Green, C. Hunt, and J. Campbell. 2008. Coastal acidification by rivers: A threat to shellfish?. *EOS, Transactions American Geophysical Union* 89(50): 513-528.

Sanyal, A., N.G. Hemming, G.N. Hanson, and W.S. Broecker. 1995. Evidence for a higher pH in the glacial ocean from boron isotopes in foraminifera. *Nature* 373: 234-236.

Sarmiento, J.L. and N. Gruber. 2006. *Ocean biogeochemical dynamics.* Princeton University Press: Princeton, NJ. 503 pp.

Schartau, A.K., S.J. Moe, L. Sandin, B. McFarland, and G.G. Raddum. 2008. Macroinvertebrate indicators of lake acidification: Analysis of monitoring data from UK, Norway and Sweden. *Aquatic Ecology* 42(2): 293-305.

Scheffer, M., S. Carpenter, J.A. Foley, C. Folke, and B. Walker. 2001. Catastrophic shifts in ecosystems. *Nature* 413(6856): 591-596.

Scheffer, M., J. Bascompte, W.A. Brock, V. Brovkin, S.R. Carpenter, V. Dakos, H. Held, E.H. van Nes, M. Rietkerk, and G. Sugihara. 2009. Early-warning signals for critical transitions. *Nature* 461: 53-59.

Scheraga, J.D. and A.E. Grambsch. 1998. Risks, opportunities and adaptation to climate change. *Climate Research* 10: 85-95.

Schiebel, R. 2002. Planktic foraminiferal sedimentation and the marine calcite budget. *Global Biogeochemical Cycles* 16(4): 1065.

Schneider, K. and J. Erez. 2006, The effect of carbonate chemistry on calcification and photosynthesis in the hermatypic coral *Acropora eurystoma. Limnology and Oceanography* 51(3): 1284-1293.

Seibel, B.L. and J.C. Drazen. 2007. The rate of metabolism in marine animals: Environmental constraints, ecological demands and energetic opportunities. *Philosophical Transactions of the Royal Society B* 362(1487): 2061-2078.

Seibel, B.A. and P.J. Walsh. 2001. Carbon cycle: Enhanced: Potential impacts of CO_2 injection on deep-sea biota. *Science* 294(5541): 319-320.

Seibel, B.A. and P.J. Walsh. 2003. Biological impacts of deep-sea carbon dioxide injection inferred from indices of physiological performance. *The Journal of Experimental Biology* 206: 641-650.

Seibel, B.A., E.V. Thuesen, J.J. Childress, and L.A. Gorodezky. 1997. Decline in pelagic cephalopod metabolism with habitat depth reflects differenes in locomotory efficiency. *The Biological Bulletin* 192: 262-278.

Seidel, M.P., DeGrandpre, M.D., and A.G. Dickson. 2008. A sensor for in situ indicator-based measurements of seawater pH. *Marine Chemistry* 109: 18-28.

Semesi, I.S., S. Beer, and M. Björk. 2009a. Seagrass photosynthesis controls rates of calcification and photosynthesis of calcareous macroalgae in a tropical seagrass meadow. *Marine Ecology Progress Series* 382: 41-47.

Semesi, I.S., J. Kangwe, and M. Björk. 2009b. Alterations in seawater pH and CO_2 affect calcification and photosynthesis in the tropical coralline alga, *Hydrolithon sp.* (Rhodophyta). *Estuarine, Coastal and Shelf Science.*

Send, U., R. Davis, J. Fischer, S. Imawaki, W. Kessler, C. Meinen, B. Owens, D. Roemmich, T. Rossby, D. Rudnick, J. Toole, S. Wijffels, and L. Beal. 2009. *A global boundary current circulation observing network.* Abstract. OceanObs 09 Conference. 21-25 September 2009. Venice, Italy.

Shi, D., Y. Xu, and F.M.M. Morel. 2009. Effects of the pH/pCO_2 control method in the growth medium of phytoplankton. *Biogeosciences Discussions* 6: 2415–2439.

Shi, D., Y. Xu, B.M. Hopkinson, and F.M.M. Morel. 2010. Effect of ocean acidification on iron availability to marine phytoplankton. *Science.* 327(5966): 676-679.

Sibuet, M. and K. Olu. 1998. Biogeography, biodiversity and fluid dependence of deep-sea cold-seep communities at active and passive margins. *Deep Sea Research Part II: Topical Studies in Oceanography* 45(1-3): 517-567.

Silverman, J., B. Lazar, and J. Erez. 2007. Effect of aragonite saturation, temperature, and nutrients on the community calcification rate of a coral reef. *Journal of Geophysical Research* 112: C05004. [doi:10.1029/2006JC003770].

Silverman, J., B. Lazar, L. Cao, K. Caldeira, and J. Erez. 2009. Coral reefs may start dissolving when atmospheric CO_2 doubles. *Geophysical Research Letters* 36.

Spivack, A.J., C-F. You, and H.J. Smith. 1993. Foraminiferal boron isotope ratios as a proxy for surface ocean pH over the past 21 Myr. *Nature* 363: 149-151.

Steinacher, M., F. Joos, T.L. Frölicher, G.-K. Plattner, and S.C. Doney. 2009. Imminent ocean acidification in the Arctic projected with the NCAR global coupled carbon cycle-climate model. *Biogeosciences* 6: 515-533.

Stephens, J.C. and D.W. Keith. 2008. Assessing geochemical carbon management. *Climate Change* 90(3): 217-242.

Stern, N., 2006. Stern review on the economics of climate change. Cambridge University Press: Cambridge, UK.

Stern, N. 2007. *The economics of climate change: The stern review.* Cambridge University Press: Cambridge, MA and New York, NY.

Sterner R.W. and J.J. Elser. (Eds.). 2002. *Ecological stoichiometry.* Princeton University Press: Oxford. U.K.

Talmage, S.C. and C.J. Gobler. 2009. The effects of elevated carbon dioxide concentrations on the metamorphosis, size, and survival of larval hard clams (*Mercenaria mercenaria*), bay scallops (*Argopecten irradians*), and Eastern oysters (*Crassostrea virginica*). *Limnology and Oceanography* 54: 2072-2080.

Thatje, S., K. Anger, J.A. Calcagno, G.A. Lovrich, H-O. Pörtner, and W.E. Arntz. 2005. Challenging the cold: Crabs reconquer the Antarctic. *Ecology* 86(3): 619-625.

Thibault, D. S. Roy, C.S. Wong, and J.K. Bishop. 1999. The downward flux of biogenic material in the NE subarctic Pacific: Importance of algal sinking and mesozooplankton herbivory. *Deep Sea Research Part II: Topical Studies in Oceanography* 46(11-12): 2669-2697.

Thistle, D., K.R. Carman, L. Sedlacek, P.G. Brewer, J.W. Fleeger, and J.P. Barry. 2005. Deep-ocean, sediment-dwelling animals are sensitive to sequestered carbon dioxide. *Marine Ecology Progress Series* 289: 1-4.

Thomas, D.J., J.C. Zachos, T.J. Bralower, E. Thomas, and S. Bohaty. 2002. Warming the fuel for the fire: Evidence for the thermal dissociation of methane hydrate during the Paleocene-Eocene thermal maximum. *Geology* 30(12): 1067–1070.

Thomas, H., L.S. Schiettecatte, K. Suykens, Y.J.M. Kone, E.H. Shadwick, A.E.F. Prowe, Y. Bozec, H.J.W. de Baar, and A.V. Borges. 2009. Enhanced ocean carbon storage from anaerobic alkalinity generation in coastal sediments. *Biogeosciences* 6(2): 267-274.

Tortell, P.D., G.R. DiTullio, D.M. Sigman, and F.M.M. Morel. 2002. CO_2 effects on taxonomic composition and nutrient utilization in an Equatorial Pacific phytoplankton assemblage. *Marine Ecology Progress Series* 236: 37-43.

Tsurumi, M., D.L. Mackas, F.A. Whitney, C. DiBacco, M.D. Galbraith, and C.S. Wong. 2005. Pteropods, eddies, carbon flux, and climate variability in the Alaska Gyre. *Deep Sea Research Part II: Topical Studies in Oceanography* 52(7-8): 1037-1053.

Tuler, S., J. Agyeman, P. Pinto da Silva, K.R. LoRusso, and R. Kay. 2008. Assessing vulnerabilities: Integrating information about driving forces that affect risks and resilience in fishing communities. *Human Ecology* 15(2):171-184.

Tunnicliffe, V., K.T.A. Davies, D.A. Butterfield, R.W. Embley, J.M. Rose, and W.W. Chadwick, Jr. 2009. Survival of mussels in extremely acidic waters on a submarine volcano. *Nature Geosciences* 2: 344-348.

Turley, C. 2005. The other CO_2 problem. In *openDemocracy*. [Online]. Available: http://www. acamedia.info/sciences/sciliterature/globalw/reference/carol_turley.html

Urcuyo, I.A., D.C. Bergquist, I.R. MacDonald, M. VanHorn, and C.R. Fisher. 2007. Growth and longevity of the tubeworm *Ridgeia piscesae* in the variable diffuse flow habitats of the Juan de Fuca Ridge. *Marine Ecology Progress Series* 344: 143-157.

U.S. Congressional Research Service (CRS). 2009. *Ocean acidification* (R40143; July 2, 2009). By E.H. Buck and P. Folger. [Online]. Available: http://assets.opencrs.com/rpts/R40143_20090702.pdf

U.S. Environmental Protection Agency (U.S. EPA). 2002. *A framework for the economic assessment of ecological benefits.* Prepared for Ecological Benefit Assessment Workgroup Social Sciences Discussion Group Science Policy Council, U.S. Environmental Protection Agency, Washington, D.C.

Van Dover, C.L. 2000. *The ecology of deep-sea hydrothermal vents.* Princeton University Press: Princeton, NJ.

van Nes, E.H. and M. Scheffer. 2004. Large species shifts triggered by small forces. *The American Naturalist* 164: 255–266.

Vogt, M., M. Steinke, S. Turner, A. Paulino, M. Meyerhöfer, U. Riebesell, C. LeQuéré, and P. Liss. 2008. Dynamics of dimethylsulphoniopropionate and dimethylsulphide under different CO_2 concentrations during a mesocosm experiment. *Biogeosciences* 5: 407-419.

von Winterfeldt, D. and W. Edwards. 1986. *Decision Analysis and Behavioral Research.* Cambridge University Press: Cambridge, MA.

Walter, L.M. and J.W. Morse. 1984. Reactive surface area of skeletal carbonates during dissolution: Effect of grain size. *Journal of Sedimentary Petrology* 54: 1081-1090.

Watson, S.A., P.C. Southgate, P.A. Tyler, and L.S. Peck. 2009. Early Larval Development of the Sydney Rock Oyster *Saccostrea glomerata* Under Near-Future Predictions of CO_2-Driven Ocean Acidification. *Journal of Shellfish Research* 28(3):431-437.

Webster, N.S., L.D. Smith, A.J. Heyward, J.E.M. Watts, R.I. Webb, L.L. Blackall, and A.P. Negri. 2004. Metamorphosis of a scleractinian coral in response to microbial biofilms. *Applied and Environmental Microbiology* 70: 1213-1221.

Weitzman, M.L. 2007. A review of the stern review on the economics of climate change. *Journal of Economic Literature* 45(3): 703-724.

Welch, C. 2009. *Oysters in deep trouble: Is Pacific Ocean's chemistry killing sea life?* The Seattle Times, June 14, 2009. [Online]. Available: http://seattletimes.nwsource.com/html/localnews/2009336458_oysters14m.html [Accessed December 2, 2009]

West, J.M., S.H. Julius, P. Kareiva, C. Enquist, J.J. Lawler, B. Petersen, A.E. Johnson, and M.R. Shaw. 2009. U.S. natural resources and climate change: Concepts and approaches for management adaptation. *Environmental Management* 44(6): 1001-1021.

Wigley, T.M.L. 2006. A combined mitigation/geoengineering approach to climate stabilization. *Science* 314(5798): 452-454.

Williams, E.A., A. Craigie, A. Yeates, and S.M. Degnan. 2008. Articulated coralline algae of the genus *Amphiroa* are highly effective natural inducers of settlement in the tropical abalone. *Haliotis asinina Biological Bulletin* 215: 98-107.

Wingenter, O.W., K.B. Haase, M. Zeigler, D.R. Blake, F.S. Rowland, B.C. Sive, A. Paulino, R. Thyrhaug, A. Larsen, K. Schulz, M. Meyerhöfer, and U. Riebesell. 2007. Unexpected consequences of increasing CO_2 and ocean acidity on marine production of DMS and CH2ClI: Potential climate impacts. *Geophysical Research Letters* 34: L05710. [doi: 10.1029/2006GL028139]

Wolf-Gladrow, D.A., U. Riebesell, S. Burkhardt, and J. Bijma. 1999. Direct effects of CO_2 concentration on growth and isotopic composition of marine plankton. *T ellus* 51B: 461–476.

Wood, H.L., J.I. Spicer, and S. Widdicombe. 2008. Ocean acidification may increase calcification rates, but at a cost. *Proceedings of Biological Sciences/ The Royal Society* 275: 1767-1773.

Worm, B., R. Hilborn, J.K. Baum, T.A. Branch, J.S. Collie, C. Costello, M.J. Fogarty, E.A. Fulton, J.A. Hutchings, S. Jennings, O.P. Jensen, H.K. Lotze, P.M. Mace, T.R. McClanahan, C. Minto, S.R. Palumbi, A.M. Parma, D. Ricard, A.A. Rosenberg, R. Watson, and D. Zeller. 2009. Rebuilding global fisheries. *Science* 325: 578-585. [DOI: 10.1126/science.1173146]

Wynne, S.P. and I.M. Côté. 2007. Effects of habitat quality and fishing on Caribbean spotted spiny lobster populations. *Journal of Applied Ecology* 44(3): 488-494.

Xu, Y., T.M. Wahlund, L. Feng, Y. Shaked, and F.M.M. Morel. 2006. A novel alkaline phosphatase in the coccolithophore *emiliania huxleyi* (prymnesiophyceae) and its regulation by phosphorus. *Journal of Phycology* 42(4): 835-844.

Yamamoto-Kawai, M, F.A. McLaughlin, E.C. Carmack, S. Nishino, and K. Shimada. 2009. Aragonite undersaturation in the Arctic Ocean: Effects of ocean acidification and sea ice melt. *Science.* 326: 1098-1100.

Yates, K.K. and R.B. Halley. 2006. CO_3^{2-} concentration and pCO_2 thresholds for calcification and dissolution on the Molokai reef flat, Hawaii. *Biogeosciences Discussions* 3:123–154.

Zachos, J.C., M. Pagani, L. Sloan, E. Thomas, and K. Billups. 2001. Trends, rhythms, and abberations in global climate 65 Ma to Present. *Science* 292: 686-693.

Zachos, J.C., U. Rohl, S.A. Schellenberg, A. Sluijs, D.A. Hodell, D.C. Kelly, E. Thomas, M. Nicolo, I. Raffi, L.J. Lourens, H. McCarren, D. Kroon. 2005. Rapid acidification of the ocean during the Paleocene-Eocene thermal maximum. *Science* 308: 1611-1615.

Zeebe, R.E. and D. Wolf-Gladrow. 2001. CO_2 in seawater: Equlibrium, kinetics, isotopes. *Elsevier Oceanography Series* 65: Elsevier Science, B.V.: Amsterdam. 346 p.

Zeebe, R.E. and J.C. Zachos. 2007. Reversed deep-sea carbonate ion basin gradient during Paleocene-Eocene thermal maximum. *Paleoceanography* 22: PA3201.

Zimmerman, R.C., D.C. Kohrs, D.L. Steller, and R.S. Alberte. 1997. Impacts of CO_2 enrichment on productivity and light requirements of eelgrass. *Plant Physiology* 115: 599-607.

Zondervan, I., B. Rost, and U. Riebesell. 2002. Effect of CO_2 concentration on the PIC/POC ratio in the coccolithophore *Emiliania huxleyi* grown under light-limiting conditions and different daylengths. *Journal of Experimental Marine Biology and Ecology* 272: 55-70.

Appendixes

A

Committee and
Staff Biographies

COMMITTEE

François M.M. Morel, *Chair,* is the Albert G. Blanke Professor of Geosciences and Director of the Center for Environmental BioInorganic Chemistry at Princeton University. He earned his Ph.D. in engineering sciences from the California Institute of Technology in 1971. Dr. Morel's research is focused on trace metal biogeochemistry, particularly the role of trace metals in the growth and activity of marine phytoplankton. One of his current projects is on the effects of decreasing pH on key chemical and biological processes such as the precipitation of calcium carbonate and the availability of major and trace nutrients. He is a fellow of the Geochemistry Society and the American Geophysical Union and is on the editorial board of several journals. Dr. Morel has served on three previous NRC committees, and was recently elected to the National Academy of Sciences.

David Archer is a professor in the Department of Geophysical Sciences at the University of Chicago. He earned his Ph.D. in oceanography from the University of Washington in 1990. He has worked on a wide range of topics pertaining to the global carbon cycle and its relation to global climate, with special focus on ocean sedimentary processes such as $CaCO_3$ dissolution and methane hydrate formation, and their impact on the evolution of atmospheric CO_2. He previously served on the NRC Organizing Committee for the First Annual Symposium on Japanese-American Frontiers of Science.

James P. Barry is a senior scientist at the Monterey Bay Aquarium Research Institute. He earned a Ph.D. in oceanography from the Scripps Institution of Oceanography in 1988. His research focuses on deep-sea biology and ecology, biological oceanography, the biology and ecology of chemosynthetic communities, climate change and marine ecosystems, polar ecology, and the biology of a high-CO_2 ocean. He is currently a member of the National Oceanographic Partnership Program's Science Advisory Panel on Investigations of Chemosynthetic Communities on the Lower Continental Slope of the Gulf of Mexico.

Garry D. Brewer is the Frederick K. Weyerhaeuser Professor of Resource Policy and Management at the Yale University School of Management. He earned his Ph.D. in political science from Yale University in 1970. Dr. Brewer is a policy scientist with broad expertise in natural resource and environmental management. Dr. Brewer has served on numerous NRC boards and committees, including chairing the Panel on Social and Behavioral Science Research Priorities for Environmental Decision Making as well as the Panel on Strategies and Methods for Climate-Related Decision Support. He was also a member of the Board on Ocean Sciences and Policy from 1983-85 and then continued from 1985-87 as a member of the Ocean Studies Board.

Jorge E. Corredor is a professor of chemical oceanography at the University of Puerto Rico at Mayagüez in their Department of Marine Sciences. Dr. Corredor earned a Ph.D. in biological oceanography from the University of Miami and a M.S. in biochemistry from the University of Wisconsin, Madison supported by Fulbright-Hays and IOC-UNESCO fellowships. He is currently researching the biogeochemistry and genomics of carbon flux in the Caribbean as forced by large river plumes and meso-scale eddies. He is also working on the establishment of an ocean observing system in the Caribbean region. Dr. Corredor is currently a member of the Ocean Studies Board.

Scott C. Doney is senior scientist in the Department of Marine Chemistry and Geochemistry at the Woods Hole Oceanographic Institution. Dr. Doney earned a Ph.D. in chemical oceanography from the Massachusetts Institute of Technology and Woods Hole Oceanographic Institution Joint Program in 1991. His research focuses on marine biogeochemistry and ecosystem dynamics, climate change, ocean acidification, and the global carbon cycle. Dr. Doney is also the chair of the Scientific Steering Committee of the Ocean Carbon and Biogeochemistry Program.

Victoria J. Fabry is a professor of biology in the Department of Biological

Sciences at California State University, San Marcos. Dr. Fabry earned a Ph.D. in biology from the University of California, Santa Barbara in 1988. Her current research focuses on the sensitivity of calcareous organisms and marine ecosystems to elevated carbon dioxide and ocean acidification, and the dissolution kinetics of biogenic calcium carbonates in the upper ocean. In 2004, Dr. Fabry presented testimony to the U.S. Senate Committee on Commerce, Science, and Transportation on the "Impacts of Anthropogenic CO_2 on Coral Reefs and Other Marine Calcifiers."

Gretchen E. Hofmann is a professor in the Department of Ecology, Evolution, and Marine Biology at the University of California, Santa Barbara. Dr. Hofmann earned a Ph.D. in Environmental, Population, and Organismal Biology from the University of Colorado in 1992. Her research focuses on the effects of climate and climate change on the performance of marine species, specifically on the impact on marine organisms of rising atmospheric CO_2 concentrations via global warming and ocean acidification. She served on the NRC Committee on the National Ecological Observatory Network.

Daniel S. Holland is a Research Scientist at the Gulf of Maine Research Institute. He was awarded his Ph.D. in environmental and natural resource economics from the University of Rhode Island in 1998. Dr. Holland's research is focused on the design and evaluation of fishery management tools and strategies that will lead to profitable and sustainable fisheries and a healthy marine ecosystem. His research methods include bioeconomic simulation modeling, econometric analysis, experimental economics, and qualitative policy analysis. He actively participates in the development of fishery policy by working with fishery stakeholders and managers to develop and evaluate policy. He is also the associate editor of *Marine Resource Economics*.

Joan A. Kleypas is a Scientist III at the National Center for Atmospheric Research. Dr. Kleypas earned a Ph.D. in Tropical Marine Studies from James Cook University, Australia in 1991. Her research focuses on how coral reefs and other marine ecosystems are affected by environmental changes associated with global climate change, such as increases in sea surface temperature and ocean acidification. Dr. Kleypas has testified at three separate U.S. Congressional hearings regarding the effects of climate change on marine ecosystems.

Frank J. Millero is a professor of marine and physical chemistry at the University of Miami Rosenstiel School of Marine and Atmospheric Science. Dr. Millero earned a Ph.D. in physical chemistry from Carnegie-

Mellon University in 1965. His general research interest is in the application of physical chemical principles to natural waters to understand how ionic interactions affect the thermodynamics and kinetics of processes occurring in the oceans. He is presently involved in studies synthesizing the global CO_2 cycle in the world oceans, including an understanding of the flux of fossil fuel CO_2 into the deep ocean. He is also interested in the role of iron as a plant nutrient and its effect on the flux of CO_2 to the deep ocean. He is a former member of the Ocean Studies Board and has served on two previous NRC committees.

Ulf Riebesell is the head of biological oceanography at the Leibniz Institute of Marine Sciences in Kiel, Germany. Dr. Riebesell earned a Ph.D. in biological oceanography from the University of Bremen, Germany in 1991. His research focuses on the sensitivity of marine organisms and ecosystems to ocean change (e.g., ocean acidification, ocean warming, changing redox conditions), the oceanic carbon cycle, the stoichiometry of marine elemental cycles, biomineralization, the biogeochemistry of stable isotopes, and paleoproxy-calibrations. He has organized and participated in numerous international conferences on ocean acidification.

STAFF

Susan Roberts became the director of the Ocean Studies Board in April 2004. Dr. Roberts received her Ph.D. in marine biology from the Scripps Institution of Oceanography. She worked as a postdoctoral researcher at the University of California, Berkeley and as a senior staff fellow at the National Institutes of Health. Dr. Roberts' past research experience has included fish muscle physiology and biochemistry, marine bacterial symbioses, and developmental cell biology. She has directed a number of studies for the Ocean Studies Board including *Nonnative Oysters in the Chesapeake Bay* (2004); *Decline of the Steller Sea Lion in Alaskan Waters: Untangling Food Webs and Fishing Nets* (2003); *Effects of Trawling & Dredging on Seafloor Habitat* (2002); *Marine Protected Areas: Tools for Sustaining Ocean Ecosystems* (2001); *Under the Weather: Climate, Ecosystems, and Infectious Disease* (2001); *Bridging Boundaries Through Regional Marine Research* (2000); and *From Monsoons to Microbes: Understanding the Ocean's Role in Human Health* (1999). Dr. Roberts specializes in the science and management of living marine resources.

Susan Park was a senior program officer with the Ocean Studies Board until the end of 2009. She received her Ph.D. in oceanography from the University of Delaware in 2004. Susan was a Christine Mirzayan Science and Technology Graduate Policy Fellow with the Ocean Studies Board in

2002 and joined the staff in 2006. She has worked on several reports with the National Academies, including *Nonnative Oysters in the Chesapeake Bay*, *Review of Recreational Fisheries Survey Methods*, *Dynamic Changes in Marine Ecosystems*, *A Review of the Ocean Research Priorities Plan and Implementation Strategy*, and *Tackling Marine Debris in the 21st Century*. Prior to joining the Ocean Studies Board, Susan spent time working on aquatic invasive species management with the Massachusetts Office of Coastal Zone Management and the Northeast Aquatic Nuisance Species Panel. She is currently Assistant Director for Research at Virginia Sea Grant.

Kathryn Hughes is a program officer with the Board on Chemical Sciences and Technology. Prior to joining the NRC staff, Kathryn was a Science Policy Fellow with the American Chemical Society. She received her Ph.D. in Analytical Chemistry from the University of Michigan, and holds a bachelors degree from Carleton College.

Heather Chiarello is a senior program assistant with the Ocean Studies Board. She graduated Magna Cum Laude from Central Michigan University in 2007 with a B.S. in political science with a concentration in public administration. Heather joined the National Academies in July 2008.

B

Acronyms

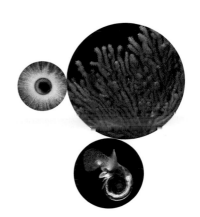

BATS	Bermuda Atlantic Time-Series
BCO-DMO	Biological and Chemical Oceanography Data Management Office
BIOACID	Biological Impacts of Ocean Acidification
CBA	Cost-benefit Analysis
CCHDO	CLIVAR and Carbon Hydrographic Data Office
CDIAC	Carbon Dioxide Information Analysis Center
CLIVAR	Climate VARiability and Predictability
CRS	U.S. Congressional Research Service
DMS	Dimethylsulfide
EMAP	Environmental Monitoring and Assessment Program
EPOCA	European Project on Ocean Acidification
FOARAM	Federal Ocean Acidification Research And Monitoring [Act of 2009]
FOCE	Free-ocean CO_2 Experiment
GCRA	Global Change Research Act
HOT	Hawaii Ocean Time-Series
ICES	International Council for the Exploration of the Sea

IGBP	International Geosphere-Biosphere Programme
IMBER	Integrated Marine Biogeochemistry and Ecosystem Research
IOC	Intergovernmental Oceanographic Commission
IOOS	Integrated Ocean Observing System
IPCC	Intergovernmental Panel on Climate Change
IWG	Interagency Working Group
JSOST	Joint Subcommittee on Ocean Science and Technology
LTER	Long-term Ecological Research
MBARI	Monterey Bay Aquarium Research Institute
MOU	Memoranda of Understanding
MSY	Maximum Sustainable Yield
NASA	National Aeronautics and Space Administration
NOAA	National Oceanic and Atmospheric Administration
NPV	Net Present Value
NRC	National Research Council
NSF	National Science Foundation
OCB	Ocean Carbon and Biogeochemistry [Program]
Ocean SITES	Ocean Sustained Interdisciplinary Time-Series Environment observation System
ODP	Ocean Drilling Program
OOI	Ocean Observatories Initiative
OSB	Ocean Studies Board
PETM	Paleocene-Eocene Thermal Maximum
PIC	Particulate Inorganic Carbon
PICES	North Pacific Marine Science Organization
POC	Particulate Organic Carbon
RISA	NOAA Regional Integrated Sciences and Assessments
SARP	NOAA Sectoral Applications Research Program
SCOR	Scientific Committee on Oceanic Research
SOLAS	Surface Ocean Lower Atmosphere Study
SRES	Special Report on Emissions Scenarios
SSS	Sea Surface Salinity
TOGA	Tropical Ocean Global Atmosphere

U.S. EPA	United States Environmental Protection Agency
U.S. GCRP	United States Global Change Research Program
U.S. GLOBEC	United States Global Ocean Ecosystems Dynamics
USGS	United States Geological Survey
U.S. JGOFS	United States Joint Global Ocean Flux Study
WCRP	World Climate Research Programme
WDC-MARE	World Data Center for Marine Environmental Sciences
WOCE	World Ocean Circulation Experiment

C

The Effect of Ocean Acidification on Calcification in Calcifying Algae, Corals, and Carbonate-dominated Systems

This appendix serves as an example of the wide variety of experimental studies on the effects of ocean acidification on calcifying marine organisms. We focus here on calcifying algae, corals, and carbonate-dominated systems, because more studies have been conducted on this collective group than on others. This table lists only those studies published through 2009 that used realistic carbonate chemistry manipulations; i.e., those that were consistent with projected changes in the carbonate chemistry of seawater due to natural forcing. Note that pCO_2 is reported both in units of parts per million (ppm) and microatmospheres (µatm); the two units can be considered essentially equivalent.

Organism/ System	Summary of findings	Reference
Calcifying Algae		
Crustose coralline algae (unidentified species)	**Manipulation**: Acid addition **Duration**: 7 weeks **Design**: Outdoor continuous-flow mesocosms: control at ambient reef pCO_2 (average 380 ppm), others manipulated to ambient + 365 ppm. Recruitment and growth of crustose coralline algae were measured on clear acrylic cylinders after 7 weeks in control and manipulated flumes. **Results**: Under high CO_2 conditions, CCA recruitment rate decreased by 78% and percentage cover decreased 92% relative to ambient; non-calcifying algae percent cover increased by 52% relative to ambient.	Kuffner et al., 2008
Rhodoliths of mixed crustose coralline algae including *Lithophyllum* cf. *pallescens*, *Hydrolithon* sp. and *Porolithon* sp.	**Manipulation**: Acid addition **Duration**: 9 months **Design**: Outdoor continuous-flow mesocosms: control at ambient reef pCO_2 (average 380 ppm), others manipulated to ambient + 365 ppm. Rhodolith growth was measured with buoyant weighing. **Results**: Rhodolith growth in control mesocosms was 250% lower than those in acidified mesocosms; that is, they experienced net dissolution.	Jokiel et al., 2008
Porolithon onkodes	**Manipulation**: Bubbled CO_2 **Duration**: 8 weeks **Design**: Algae placed in flow-through aquaria: 2 temperatures: 25–26°C and 28–29°C; 3 pH levels: 8. 0–8.4 (control) 7.85–7.95 and 7.60–7.70. **Results**: *P. onkodes* calcification rate in low pH treatment was 130% less (25–26°C) and 190% less (28–29°C) than in control (i.e., net dissolution).	Anthony et al., 2008

Organism/ System	Summary of findings	Reference
Calcareous epibionts on seagrasses (*Hydrolithon boreale, H. cruciatum, H. farinosum, Pneophyllum confervicola, P. fragile* and *P. zonale*)	**Manipulation**: Bubbled CO_2 and field observations **Duration**: 2 weeks **Design**: In field, calcium carbonate mass on seagrass blades was measured across a natural pH gradient. In lab, seagrass blades with 50-70% cover of crustose coralline algae were collected from the field and placed in aquaria of pH = 8.1 (control) or pH = 7.0. Coralline algal cover was estimated before and after treatments. **Results**: In field, coralline algal cover was highly correlated with pH, decreasing rapidly below pH = 7.8 and absent at pH = 7.0; in lab experiment, coralline algae were completely dissolved after two weeks at a pH of 7.0, whereas control samples showed no discernable change.	Martin et al., 2008
Rhodoliths of *Hydrolithon* sp.	**Manipulation**: Both acid/base addition and bubbled CO_2 **Duration**: 5 days **Design**: Acid/base additions used to alter pH to multiple levels (7.6, 7.8, 8.2, 8.6, 9.0, 9.4 and 9.8; control was 8.1); CO_2 bubbling used to alter pH and DIC to 7.8. **Results**: Calcification rate was positively correlated with pH in both light and dark experiments; decreasing the pH to 7.8 with CO_2 bubbling lowered calcification by 20%.	Semesi et al., 2009a
Hydrolithon sp. *Mesophyllym* sp. *Halimeda renschii*	**Manipulation**: Drawdown of CO_2 by seagrass photosynthesis **Duration**: 2.5 hours **Design**: In situ open-bottom incubation cylinders; pH and algal calcification rates measured in presence or absence of seagrasses. **Results**: Seagrass photosynthesis caused pH to increases from 8.3–8.4 to 8.6–8.9 after 2.5 hours; calcification rates increased > 5x for *Hydrolithon* sp., and 1.6x for *Mesophyllum* sp. and *Halimeda* sp.	Semesi et al., 2009b

Organism/ System	Summary of findings	Reference
Lithophyllum cabiochae	**Manipulation**: Bubbled CO_2 **Duration**: 1 year **Design**: Algae were maintained in aquaria at ambient or elevated temperature (+3°C) and at ambient (~400 ppm) or elevated pCO_2 (~700 ppm). **Results**: No clear pattern of reduced calcification at elevated pCO_2 alone, but combination of elevated pCO_2 and temperature led to high rates of necroses and death. The dissolution of dead algal thalli at elevated pCO_2 was 2–4x higher than under ambient pCO_2.	Martin and Gattuso, 2009
Corallina sessilis	**Manipulation:** Bubbled CO_2 **Duration:** 30 days **Design:** Controlled laboratory experiments to investigate the interactive effects of pCO_2 and UV radiation on growth, photosynthesis, and calcification. 2 pCO_2 levels (280 and 1000 ppmv), combined with 3 light conditions: PAR alone (solar radiation wavelengths > 395 nm); PAR+UVA (> 320 nm); PAR+UVA+UVB (> 295 nm). **Results:** Under PAR alone, elevated pCO_2 decreased net photosynthetic rate by 29.3%, and calcification rate by 25.6% relative to low pCO_2. Elevated pCO_2 exacerbated the effects of ultraviolet radiation in inhibiting rates of growth (from 13% to 47%), photosynthesis (from 6% to 20%), and calcification (from 3% to 8%). The authors suggest that the decrease in calcification in *C. sessilis* at higher pCO_2 levels increases its susceptibility to damage by UVB radiation.	Gao and Zheng, 2009
Halimeda incrassata (green alga) and *Neogoniolithon spp.* (coralline red alga)	**Manipulation:** CO_2 bubbling **Duration:** 60 days **Design:** Controlled laboratory experiment to examine changes in calcification under Ω_{arag} = 3.12, 2.40, 1.84, and 0.90 (approx. pCO_2 = 409, 606, 903, 2856 ppmv, respectively). SST maintained at 25°C. **Results:** Calcification rates in both species were higher at Ω_{arag} = 2.40, then declined at lower saturation states.	Ries et al., 2009

Organism/ System	Summary of findings	Reference
Corals		
Stylophora pistillata	**Manipulation**: Altered Ca^{2+} ion concentration[1] **Duration**: 2.5 hours **Design**: Controlled laboratory experiment; aragonite saturation changes from 98 to 390% were obtained by manipulating the calcium concentration. **Results**: Nonlinear increase in calcification rate as a function of aragonite saturation level.	Gattuso et al., 1998
Porites compressa	**Manipulation:** Acid addition **Duration:** 5 weeks **Design:** 760 and 3980 µatm (pH = 8.2 versus 7.2); nitrate additions as well **Results:** Corals grown in low pH water grew half as fast.	Marubini and Atkinson, 1999
Porites compressa	**Manipulation**: Acid addition **Duration**: 10 weeks **Design**: Controlled laboratory experiments: measured calcification at pCO_2 = 199 and 448 µatm, at 3 light levels. In Biosphere 2 coral mesocosm: measured calcification at pCO_2 = 186, 336, and 641 µatm. **Results**: Calcification decreased 30% from pCO_2 = 186 to 641, and 11% from pCO_2 = 336 to 641 µatm, regardless of light level.	Marubini et al., 2001
Galaxea fascicularis	**Manipulation**: Altered Ca^{2+} ion concentration while maintaining pH at 8.11–8.12; temperatures maintained at ambient temperature of collections site[1] **Duration**: Hours **Design**: Calcium additions to estimated Ω_{arag} from 3.88 (present-day) to 4.83 and 5.77; calcification rate measured with ^{14}C incorporation in skeleton. **Results**: Calcification rate increased 30–60% at Ω_{arag} = 4.83 and 50–80% at Ω_{arag} = 5.77 relative to Ω_{arag} =3.88.	Marshall and Clode, 2002

Organism/ System	Summary of findings	Reference
Stylophora pistillata	**Manipulation**: Bubbled CO_2 **Duration**: 5 weeks **Design**: 2 pCO_2 values (460 and 760 µatm) and 2 temperatures (25 and 28°C) **Results**: Calcification under normal temperature did not change in response to an increased pCO_2. Calcification decreased by 50% when temperature and pCO_2 were both elevated.	Reynaud et al., 2003
Acropora verweyi Galaxea fascicularis Pavona cactus Turbinaria reniformis	**Manipulation**: Acid/base addition **Duration**: 8 days **Design**: 2 pCO_2 values (407–416 and 857–882 µatm), 26.5°C **Results**: calcification rate in all 4 species decreased 13–18%	Marubini et al., 2003
Porites compressa + *Montipora capitata*	**Manipulation**: acid/base addition **Duration**: 1.5 hours **Design**: Corals placed in flumes, multiple summer experiments at pCO_2 = 460 and 789 µatm; multiple winter experiments at pCO_2 = 391, 526, and 781 µatm; additional experiments included additions of PO_4 and NH_4. **Results**: Summer calcification rate declined 43% with increase in pCO_2 from 460 to 789 µatm; winter rates declined 22% from 391 to 526 µatm; and 80% from 391 to 781 µatm.	Langdon and Atkinson, 2005
Acropora cervicornis	**Manipulation**: Bubbled CO_2 **Duration**: 16 weeks total **Design**: Nubbins cultured for 1 week at pCO_2=367 µatm, 2 weeks at 714–771 µatm, 1 week at 365 µatm **Results**: 60–80% reduction in calcification rate at 714–771 µatm relative to controls (357–361 µatm); note that calcification rate did not substantially recover with return to normal pCO_2 during 4th week.	Renegar and Riegl, 2005

Organism/ System	Summary of findings	Reference
Acropora eurystoma	**Manipulation**: Acid/base addition **Duration**: Hours **Design**: Separation of effects of different carbonate chemistry parameters by maintaining a) constant total inorganic carbon, b) constant pH, or c) constant CO_2; temperatures = 23.5–24.5°C **Results**: calcification rate was correlated with $[CO_3^{2-}]$: 50% decrease in calcification with 30% decrease in $[CO_3^{2-}]$; 35% decrease in calcification with increase in pCO_2 from 370 to 560 ppm.	Schneider and Erez, 2006
Porites lutea and *Fungia* sp.	**Manipulation**: Acid/base addition **Duration**: 3 hours (night-time) and 6 hours (day-time) **Design**: Coral colonies were acclimated for several months, then subjected to seawater adjusted to one of 3 Ω_{arag} levels: 1.56, 3.43, 5.18 (note that ambient Ω_{arag} was 3.43); temperature was constant at 25°C. **Results**: Both day and night calcification decreased with decreasing pH; calcification rate at 2x preindustrial CO_2 level (Ω_{arag} = 3.1) was reduced by 42% relative to preindustrial level (Ω_{arag} = 4.6).	Ohde and Hossain, 2004; Hossain and Ohde, 2006
Montipora capitata	**Manipulation**: Acid addition **Duration**: 10 months **Design**: Corals places in flumes: control at ambient reef pCO_2 (average 380 ppm), others manipulated to ambient + 365 ppm. **Results**: Calcification decreased 15–20% with a doubling of pCO_2 (380 to 380+365 ppm).	Jokiel et al., 2008
Porites astreoides (larvae/juveniles)	**Manipulation**: Acid addition **Duration**: 21–28 days **Design**: Flow-through seawater system; 3 aragonite saturation states: Ω_{arag} = 3.2 (control), 2.6 (mid), and 2.2 (low); constant temperature at 25°C **Results**: Lateral skeletal extension in larvae was positively correlated with saturation state ($P=0.007$); juveniles in mid Ω_{arag} treatment grew 45–56% slower than controls; those in low Ω_{arag} treatments grew 72–84% slower than controls.	Albright et al., 2008

Organism/ System	Summary of findings	Reference
Porites lobata *Acropora intermedia*	**Manipulation**: Bubbled CO_2 **Duration**: 8 weeks **Design**: Corals placed in flow-through aquaria: 2 temperatures: 25–26°C and 28–29°C; and 3 pH levels: 8. 0–8.4 (control) 7.85–7.95 and 7.60–7.70. **Results**: *Acropora intermedia* and *Porites lobata* calcification rates were 40% lower at low pH treatment than in control.	Anthony et al., 2008
Favia fragrum (larvae/juveniles)	**Manipulation**: Acid addition **Duration**: 8 days **Design**: Newly settled coral larvae reared in a range of Ω_{arag} from ambient (3.71) to 3 treatments (Ω_{arag} = 2.40, 1.03, 0.22); culture temperatures =25°C. **Results**: Aragonite was secreted by all corals even in undersaturated conditions; however, in Ω_{arag} = 2.40 treatment, cross-sectional area of skeletons was more than 20% less than the control, and average weight of skeletal mass was 26% less than control. Similar trends occurred in the more extreme treatments.	Cohen et al., 2009
Madracis mirabilis	**Manipulation**: Acid/base addition and bubbled CO_2 **Duration**: 2 hour incubations following 3-hour acclimation period **Design**: Separation of effects of different carbonate chemistry parameters by manipulating chemistry to reflect 6 combinations of normal, low and very low pH, with normal low and very low $[CO_3^{2-}]$; temperature maintained at 28°C **Results**: For pH/$[CO_3^{2-}]$ combinations that simulate natural ocean acidification (pCO$_2$ = 390, 875 and 1400 µatm), calcification rate was not correlated with $[CO_3^{2-}]$, but rather with $[HCO_3^-]$.	Jury et al., 2009

Organism/ System	Summary of findings	Reference
Oculina arbuscula (temperate coral)	**Manipulation:** CO_2 bubbling **Duration:** 60 days **Design:** Controlled laboratory experiment to examine changes in calcification under Ω_{arag} = 3.12, 2.40, 1.84, and 0.90 (approx. pCO_2 = 409, 606, 903, 2856 ppmv, respectively). SST maintained at 25°C. **Results:** Calcification rate remained unchanged Ω_{arag} > 1.84, then declined rapidly at Ω_{arag} = 0.90.	Ries et al., 2009
Lophelia pertussa (cold water coral)	**Manipulation:** Acid addition **Duration:** 24 hours **Design:** On-board incubations of deep-water corals at ambient pH, ambient pH – 0.15 units, and ambient pH – 0.3 units. Calcification rates measured using ^{45}Ca labeling. **Results:** Calcification rates were reduced by 30% and 56% at pH reduced by 0.15 and 0.3 units, respectively, as compared to calcification rate at ambient pH. Calcification in young polyps showed a stronger reduction than in old polyps (59% reduction versus 40% reduction, respectively).	Maier et al., 2009
Carbonate-dominated systems Gr. Bahama Banks	**Manipulation:** NA; field measurements **Duration:** Days **Design:** Measured changes in pCO_2, DIC, temperature salinity, and residence time of Bahama Banks waters. **Results:** $CaCO_3$ precipitation rate correlated with $CaCO_3$ saturation state.	Broecker and Takahashi, 1966; Broecker et al., 2001
B2 mesocosm	**Manipulation:** Acid/base and $CaCl_2$ additions and natural alkalinity draw-down **Duration:** Days to months/years (3.8 years total) **Design:** Biosphere 2 coral reef mesoscosm; time series of net community calcification measurements in relation to carbonate chemistry. **Results:** Calcification rate well correlated with saturation state; calcification rate decreased 40% between preindustrial and doubled CO_2 conditions.	Langdon et al., 2000; Langdon et al., 2003

Organism/ System	Summary of findings	Reference
Monaco mesocosm	**Manipulation**: Bubbled CO_2 **Duration**: 24-hour incubations **Design**: Coral community mesocosm subjected to continuous flow with a range of pCO_2 values (134–1813 µatm; temperature maintained at 26°C **Results**: Community calcification was reduced by 21% between preindustrial and double pCO_2 levels.	Leclercq et al., 2000
Monaco mesocosm	**Manipulation**: Bubbled CO_2 **Duration**: 9–30 days **Design**: Coral community mesocosm subjected to continuous flow with mid (647 µatm) pCO_2 for 12 weeks, low (411 µatm) for 4 weeks, and high (918 µatm) for 4 weeks; temperature maintained at 26°C **Results**: Daytime community calcification was reduced by 12% between low and high treatments.	Leclercq et al., 2002
Molokai Reef System	**Manipulation**: Natural alkalinity drawdown by organisms **Duration**: Several days **Design**: Large benthic chambers placed on reef bed; in situ carbonate chemistry, salinity, temperature, and net calcification/dissolution measured continuously. **Results**: Calcification and dissolution were linearly correlated with both CO_3^{2-} and pCO_2. Threshold pCO_2 and CO_3^{2-} values for individual substrate types showed considerable variation. Results indicate that average threshold for shift to net dissolution for Molokai reef is when $pCO_2 = 654 \pm 195$ µatm.	Yates and Halley, 2006

Organism/ System	Summary of findings	Reference
Northern Red Sea Reef	**Manipulation**: NA; field measurements **Duration**: 2 years **Design**: Eulerian measurements of carbonate system in seawater and community calcification/dissolution rates as a function of saturation state; adjusted for residence time of water. **Results**: Based on seasonal differences in calcification rate, determine that net reef calcification rate was well-correlated with precipitation rates of inorganic aragonite; projected a 55% decrease in reef calcification at 560 ppm CO_2 and 30°C relative to 280 ppm and 28°C	Silverman et al., 2007
Calcifying community dominated by *Montipora capitata*	**Manipulation**: Acid addition **Duration**: 24 hours **Design**: See Jokiel et al., 2008 and Kuffner et al. 2008. Compared Net ecosystem calcification (NEC) in coral community mesosms exposed to ambient pCO_2 (380 ppm) and 2x ambient (380+365 ppm). NEC was determined every 2 hours by accounting for changes total alkalinity in the entire system. **Results**: NEC was 3.3 mmol $CaCO_3$ m^{-2} h^{-1} under ambient and −0.04 mmol $CaCO_3$ m^{-2} h^{-1}.	Andersson et al., 2009

[1]These studies manipulated Ca^{2+} rather than the carbon system. They are included here for completeness and because they provide insights into calcification mechanisms, but the results should not be strictly interpreted as a response to ocean acidification.

D

Summary of Research Recommendations from Community-based References

Multiple documents have addressed the need for ocean acidification research, and five of these were regarded by the committee as both community-based, in that they included broad input from scientists, and forward looking, in that they made specific recommendations for research needs. The summary and recommendations from each report include:

Raven, J., K. Caldeira, H. Elderfield, O. Hoegh-Guldberg, P.S. Liss, U. Riebesell, J. Shepard, C. Turley and A.J. Watson. 2005. *Ocean acidification due to increasing atmospheric carbon dioxide*. Policy Document. *The Royal Society*, London, 60 pp.

Summary: This report, produced by the UK Royal Society's Working Group on Ocean Acidification, was the first comprehensive report on the chemical and biological impacts of ocean acidification. It provides a detailed summary of the effects of ocean acidification, and makes conclusions and recommendations for policymakers. The working group identified the following priority research areas:

• Identification of species, functional groups, and ecosystems that are most sensitive ocean acidification and the rate at which organisms can adapt to the changes
• Interaction of increased CO_2 in surface oceans with other factors such as temperature, carbon cycle, sediment processes, and the balance of reef accretion and erosion

- Feedback of increased ocean surface CO_2 on air-sea exchange of CO_2, dimethlysulphide and other gases important for climate and air quality
- Large-scale manipulation experiments on the effect of increased CO_2 on biota in the surface waters.

Kleypas, J.A., R.A. Feely, V.J. Fabry, C. Langdon, C.L. Sabine, and L.L. Robbins. 2006. *Impacts of Ocean Acidification on Coral Reefs and Other Marine Calcifiers: A Guide for Future Research,* **report of a workshop held 18-20 April 2005, St. Petersburg, FL, sponsored by NSF, NOAA, and the U.S. Geological Survey, 99 pp.**

Summary: The paper is the result of a workshop, sponsored by NSF, NOAA, and the USGS. Roughly 50 scientists participated from a wide range of disciplines. The aims of the workshop were to summarize existing knowledge on the topic of ocean acidification impacts on marine calcifiers, reach a consensus on what the most pressing scientific issues are, and identify future research strategies for addressing these issues. The report is intended as a guide to program managers and researchers toward designing research projects with the details and references needed to address the major scientific issues that should be pursued in the next 5-10 years.

- Develop protocols for the various methodologies used in seawater chemistry and calcification measurements
- Determine the calcification response to elevated CO_2 in benthic and planktonic calcifiers
- Physiological research to discriminate the various mechanisms of calcification within calcifying groups, to better understand the cross-taxa range of responses to changing seawater chemistry
- Experimental studies to determine the interactive effects of multiple variables that affect calcification and dissolution in organisms (saturation state, light, temperature, nutrients)
- Combining laboratory experiments with field studies to establish clear links between laboratory experiments and the natural environment
- Long-term monitoring of coral reef response to ocean acidification, and better accounting of calcium carbonate budgets
- Monitoring of in situ calcification and dissolution in organisms
- Incorporating ecological questions into observations and experiments; e.g., effects on organism survivorship and ecology, ecosystem functioning, etc.
- Biogeochemical and ecological modeling to improve understanding of carbonate system interactions, and to guide future sampling and experimental efforts

Fabry, V.J., C. Langdon, W.M. Balch, A.G. Dickson, R.A. Feely, B. Hales, D.A. Hutchins, J.A. Kleypas, and C.L. Sabine. 2008. *Present and Future Impacts of Ocean Acidification on Marine Ecosystems and Biogeochemical Cycles***, report of the Ocean Carbon and Biogeochemistry Scoping Workshop on Ocean Acidification Research held 9-11 October 2007, La Jolla, CA, 40 pp.**

Summary: This report is a result of the Ocean Carbon and Biogeochemistry (OCB) Scoping Workshop on Ocean Acidification Research sponsored by NSF, NOAA, NASA, and USGS. This report summarizes input from nearly 100 scientists in a comprehensive research strategy for four critical ecosystems: warm-water coral reefs, coastal margins, subtropical/tropical pelagic regions, and high latitude regions over immediate (2-5 yrs) and long-term (5-10 yrs) time scales. The key overall recommendations for research include:

- Establish a national program on ocean acidification research
- Develop new instrumentation for the autonomous measurement of CO_2 system parameters, particulate inorganic carbon (PIC), particulate organic carbon (POC), and physiological stress markers
- Standardize protocols for manipulation and measurement of seawater chemistry in experiments and for calcification and other rate measurements
- Expand existing ocean CO_2 system monitoring to include new monitoring sites/surveys in open-ocean and coastal regions, including sites considered vulnerable to ocean acidification, and sites that can be leveraged for field studies
- Establish new monitoring sites/surveys in open-ocean and coastal regions, including sites of particular interest such as the Bering Sea
- Progressively build capacity and initiate planning for mesocosm and CO_2-perturbation experiments in the field
- Build shared facilities to conduct well-controlled CO_2-manipulation experiments
- Perform global data/model synthesis to predict and quantify alterations in the ocean CO_2 system due to changes in marine calcification
- Develop regional biogeochemical models and conduct model/data intercomparison analyses
- Establish international collaborations to create a global network of CO_2 system observations and field studies relevant to ocean acidification
- Ensure that the research is designed to provide results that are useful for policy and decision making
- Initiate specific activities for education, training, and outreach

Orr, J.C., K. Caldeira, V. Fabry, J.P. Gattuso, P. Haugan, P. Lehodey, S. Pantoja, H.O. Pörtner, U. Riebesell, and T. Trull, M. Hood, E. Urban, and W. Broadgate. 2009. *Research Priorities for Ocean Acidification*, report from the Second Symposium on the Ocean in a High-CO$_2$ World, Monaco, October 6-8, 2008, convened by SCOR, UNESCO-IOC, IAEA, and IGBP, 25 pp.

Summary: The *Research Priorities Report* resulted from the 2nd symposium on *The Ocean in a High-CO$_2$ World*, held in 2008 in Monaco. The symposium was sponsored by SCOR, IOC, other international groups, and the U.S. NSF, and included 220 scientists from 32 countries to assess what is known about the impacts of ocean acidification on marine chemistry and ecosystems. The *Research Priorities Report* highlights new findings and details the research priorities identified by the symposium participants during discussion sessions on 1) perturbation experiments, 2) observation networks, and 3) scaling organism-to-ecosystem acidification effects and feedbacks on climate:

Observations

• Develop new instrumentation for autonomous measurements of CO$_2$ system parameters, particulate inorganic (PIC), particulate organic carbon (POC), and other indicators of impacts on organisms and ecosystems;
• Maintain, enhance, and extend existing long-term time series that are relevant for ocean acidification; establish new monitoring sites and repeat surveys in key areas that are likely to be vulnerable to ocean acidification;
• Develop relaxed carbon measurement methods and appropriate instrumentation that are cheaper and easier, if possible, for high-variability areas that may not need the highest measurement precision;
• Establish a high-quality ocean carbon measurement service for those unable to develop their own measurement capabilities;
• Establish international collaborations to create a data management and synthesis program for new ocean acidification data as well as data mining and archival for relevant historical data sets;
• Work on developing an ocean acidification index (e.g., a CaCO$_3$ saturation index based on a standard carbonate material);
• Initiate specific activities for education, training, and outreach.

Perturbation Experiments

• Controlled single-species laboratory experiments to look at species responses, to improve understanding of physiological mechanisms, and

to identify longer-term, multi-generational adaptation (both physiological and behavioral);
• Microcosms and mesocosms to elucidate community responses and to validate and up-scale single-species responses;
• Natural perturbation studies from CO_2 venting sites and naturally low pH regions such as upwelling regions, which provide insights to ecosystem responses, long-term effects, and adaptation mechanisms in low-pH environments;
• Manipulative field experiments; and
• Mining the paleo-record to develop and test hypotheses.

Scaling from organism to ecosystems

• Determine which ecosystems are at the greatest risk from ocean acidification and which of these are most important
• Determine ecological tipping points that can be defined in terms of pH or carbonate ion concentration
• Determine which physiological processes are most important to the scaling issue
• Determine how impacts of ocean acidification scale from life stages and individuals to populations, ecosystems and biodiversity; assess biological interactions and fluxes across trophic levels
• Determine impacts of ocean acidification on fisheries, food production, and other ecosystem services; Increase integrated research involving physiologists, ecologists and fisheries scientists to determine food web responses
• Investigate how ecosystem-ecosystem linkages will be affected by ocean acidification (including pelagic-benthic linkages)
• Investigate the potential for behavioral adaptation (e.g., migration and avoidance) to ocean acidification?

Joint, I., D.M. Karl, S.C. Doney, E.V. Armbrust, W. Balch, M. Berman, C. Bowler, M. Church, A. Dickson, J. Heidelberg, D. Iglesias-Rodriguez, D. Kirchman, Z. Kolber, R. Letelier, C. Lupp, S. Maberly, S. Park, J. Raven, D.J. Repeta, U. Riebesell, G. Steward, P. Tortell, R.E. Zeebe and J.P. Zehr. 2009. *Consequences of high CO_2 and ocean acidification for microbes in the global ocean,* Report of expert meeting at U. Hawaii, 24-26 February 2009 organized by Plymouth Marine Laboratory and Center for Microbial Oceanography Research and Education, 23 pp.

Summary: This report is a summary of a workshop attended by 24 scientists, predominantly marine microbial oceanographers, at the Center for

Microbial Oceanography and Education (University of Hawaii) in February 2009. The goal of the workshop was to assess the consequences of higher CO_2 and lower pH for marine microbe and to define high-priority research questions. The report identifies ten important questions related to the effects of acidification on marine microbes, and attempts to indicate urgency and the likely scale of investment that will be required. The top ten priorities are:

• Agreement on best methods to manipulate seawater chemistry for biological incubations. Can specific changes/biological responses be isolated (e.g., pH versus pCO_2 vs. carbonate ion)?
• Basic studies on how microbial physiology responds to pH change (e.g., internal cellular controls on pH). This may require development of new techniques (e.g., single cell manipulation).
• Accessing genomic information of how natural populations respond to pH change using metagenomic and metatranscriptomics approaches.
• Single species studies on CO_2 and pH sensitivity across major groups (i.e., calcifiers, photosynthesizers, nitrogen-fixers, and heterotrophic bacteria).
• Comparison of ocean zones of high respiration (high natural pCO_2) and tropical versus polar (cold water seas).
• Freshwater and estuarine microbes accommodate frequent and rapid natural pH change. Are marine microbes less adaptable to pH change?
• What are the time scales of adaptation (evolution) to higher CO_2 and lower pH and can this be demonstrated in laboratory cultures?
• How will complex natural assemblages respond to higher CO_2 and lower pH over time scales of years to decades?
• How will open ocean ecosystems structure respond to higher CO_2 and lower pH? Can mesocosm experiments be extended to the open ocean?
• Mesoscale CO_2-enrichment experiments (similar to iron-enrichment studies).